Searching for Water in the Universe

Thérèse Encrenaz

Searching for Water in the Universe

Published in association with
Praxis Publishing
Chichester, UK

Dr Thérèse Encrenaz
Laboratoire d'Etudes Spatiales et d'Instrumentation en Astrophysique (LESIA)
Paris Observatory
Meudon
France

Original French edition: *A la recherche de l'eau dans l'univers*
Published © Éditions Belin 2004
Ouvrage publié avec le concours du Ministère français chargé de la culture - Centre national du livre
This work has been published with the help of the French Ministère de la Culture - Centre National du Livre

Translator: Bob Mizon, 38 The Vineries, Colehill, Wimborne, Dorset, UK

SPRINGER-PRAXIS BOOKS IN POPULAR ASTRONOMY
SUBJECT *ADVISORY EDITOR*: John Mason B.Sc., M.Sc., Ph.D.

ISBN 10: 0-387-34174-9 Springer Berlin Heidelberg New York
ISBN 13: 978-0-387-34174-3

Springer is a part of Springer Science + Business Media (*springeronline.com*)

Library of Congress Control Number: 2006926438

Cover design: Jim Wilkie
Copy editing: R. A. Marriott
Typesetting: BookEns Ltd, Royston, Herts., UK

Printed in Germany on acid-free paper

To Pierre

I would like to thank Belin Publishing and Praxis Publishing for their valued and efficient support in the realisation of this book, Fabienne Casoli for reading the original text in the French edition, and Bob Mizon for his excellent translation into English. I am also grateful to Pierre Cox, who, with his specialist understanding of the question of water in the interstellar medium, has helped guide me beyond the bounds of the solar system. Finally, I wish to thank all those colleagues who gave me access to their documents, so useful to the fabric of this text.

Contents

Introduction
Why water?

Interestingly, water is present on our planet in three different states: vapour, liquid and solid. In its liquid form, it has played an essential part in the appearance, development and maintenance of terrestrial life. What is its role elsewhere in the Universe? In its gaseous and solid forms, water is omnipresent: in the most distant galaxies, among the stars, in the Sun, in its planets and their satellites and ring systems, and in comets.

Is there extraterrestrial life? We still await the answer, and the search for liquid water is an indispensable aspect of that answer.

If the Earth's oceans did not exist, we would not be here to ask why. Nowadays, everyone understands the essential and undeniable role that liquid water has played in the emergence, development and maintenance of life on Earth. Liquid water is by far the major constituent of the mass of living organisms, be they animals or plants. And, looking to the future, the availability of our planet's reserves of fresh water is an ever more serious issue for us all, as population numbers soar. Access to that water will present a great challenge during the century which has just begun.

We know that the abyssal oceans gave rise to the first signs of life on Earth, but what is less well known is that those oceans have ensured the stability of our habitable planet, with its temperate climates. Since the earliest times, the Earth's temperature has remained relatively stable: we have not had to suffer the kind of runaway greenhouse effect which characterises Venus, leading to scorching surface temperatures of 730 K (about 450° C); neither does the Earth's surface resemble the freezing deserts of Mars, with their average temperature of 230 K (–43° C).

The greenhouse effect

The greenhouse effect is a mechanism causing warming of the surface and lower atmosphere of the Earth or other planet. The surface, heated by solar radiation (especially in the visible part of the spectrum at wavelengths of 0.4–0.8 μm), reaches a stable temperature regulated by the fraction of solar radiation reflected, known as the albedo (of the order of 0.3 for the Earth), the rest being absorbed and converted into thermal energy. For Earth, this equilibrium temperature is 255 K (–18° C). At this temperature, a black body (absorbing all incident electromagnetic energy) emits mainly in the infrared ($\lambda > 1$ μm). In the case of the Earth, the infrared radiation emitted from the surface is absorbed by two atmospheric gases: water vapour and carbon dioxide. The absorption of surface radiation by the lower atmosphere in turn contributes to the warming of the surface, and the process is amplified. This is known as the greenhouse effect, so called because it is analogous to the mechanism

---

Clouds above the Pacific. An image taken from the International Space Station on 21 July 2003.

whereby a greenhouse is heated, its glass playing the role of the lower atmosphere and letting through the visible radiation but blocking the infrared. On Earth, this involves a heating effect of 33° C – a modest value kept constant by a self-regulatory mechanism involving the oceans. The phenomenon is less marked on Mars (4° C), but it was undoubtedly much more important in the past. On Venus, the effect is dramatic. The surface heated to a temperature of 730 K (more than 450° C), showing how the mechanism can run wild if no regulation is present. This illustrates the threat to the Earth's climate posed by increased quantities of carbon dioxide, if humans continue to produce it at current rates.

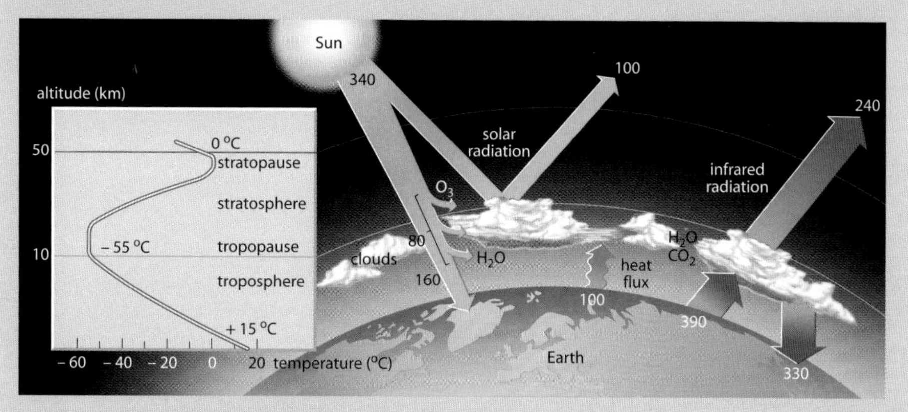

Figure 1. The mechanism of the greenhouse effect. Some of the Sun's radiation reaches the surface and warms it. The surface then emits infrared radiation which is absorbed by infrared-active gases in the lower atmosphere (CO_2, H_2O). The lower atmosphere, thus warmed, re-emits radiation towards the surface and further warms it, amplifying the phenomenon. The numbers indicate the radiation budget in W/m^2. (From S. Jousseaume, *Climat d'Hier à Demain*, CNRS Editions-CEA, 1993.)

LIFE ON EARTH...

The first thing to note is that the water molecule, a very minor constituent (<0·03%) of the general mass of our planet, is unique in that it can exist in its three states of vapour, liquid and solid. The liquid water of the oceans, lakes and other water features is by far the most abundant of the three. Water in vaporous form is the least abundant. Atmospheric pressure due to water vapour represents less than one-hundredth of total atmospheric pressure, given the dominance of nitrogen and oxygen. But appearances are deceptive: if all the water of the oceans and glaciers were transformed into vapour, then we would experience, at the surface, an increase in atmospheric pressure of more than a hundredfold!

It seems that the Earth has been fortunate enough to hold on to its initial reserves of water, while nearby planets have not, as will be seen in chapter 7. It is

Figure 2. The Pacific Ocean. The presence of oceans has led to the emergence of life on Earth, and to the temperate climate which ensures the habitability of our planet.

to those reserves that we owe the fact that the carbon dioxide (CO_2) which was so abundant in the beginning has been dissolved into the oceans to form calcium carbonate ($CaCO_3$). Freed of its load of carbon dioxide, a major greenhouse gas (as is water vapour), the Earth has been spared the worst of the greenhouse effect and its temperature has remained relatively unchanged.

We know that the Earth was formed some 4·6 billion years ago, together with its sister planets, by a process of accretion involving smaller bodies, or planetesimals. Life seems to have emerged very early on in the history of this planet, after about a billion years or so. How did this occur? This is a question which remains to be answered. It is, however, clear that the presence of liquid water played a determining role in the process, and the first known life forms appeared in the ocean depths. Although water and the presence of prebiotic molecules (i.e. complex organic ones such as amino acids) are not in themselves

Figure 3. The Earth is the only planet in the solar system where water is present in all three phases: vapour, liquid and solid. Its liquid reservoir, in the oceans, is by far the most abundant, while water vapour is the least abundant.

the precursors of life, they do appear to be essential to the emergence of life as we know it.

The obvious question arises: if the oceans played such a vital role in the development of life on Earth, could the same thing have happened elsewhere in the solar system, or beyond? Well over a hundred planets have already been discovered outside the solar system, in the nearby Universe. These are known as extrasolar planets or *exoplanets*, and new and more powerful search methods will undoubtedly reveal many hundreds more in the near future. Where must we look, among all these exoplanets, for those which may harbour life? Using the analogy provided by our experience of life on Earth, likely candidates will be those planets which resemble ours, and may well possess a liquid ocean on their surfaces. Such bodies are called *exoearths* (sometimes written *exoEarths*), signifying exoplanets which may have a solid surface, and an atmospheric mass much smaller than their total mass, as is the case with the terrestrial planets of the solar system.

And so to our more general question: does water exist out there in the Universe? For some years now we have known the answer. It is yes. Water is everywhere in the Universe, in the most distant galaxies, among stars young and old, in the Sun and nearly all its family: planets and their satellites and ring systems, and comets. Our certainty that there is water in the distant Universe is due largely to the European Infrared Space Observatory (ISO), launched in 1995 and operational in Earth orbit until 1998. Why use an orbiting probe to investigate the presence of water in the far Universe? Simply because, in order to detect water in the cosmos, it is necessary to be outside the atmosphere, which contains so much of it.

Figure 4. Atmospheric movements above the island of Guadalupe, off Mexico's Pacific coast.

Water in the Earth's atmosphere

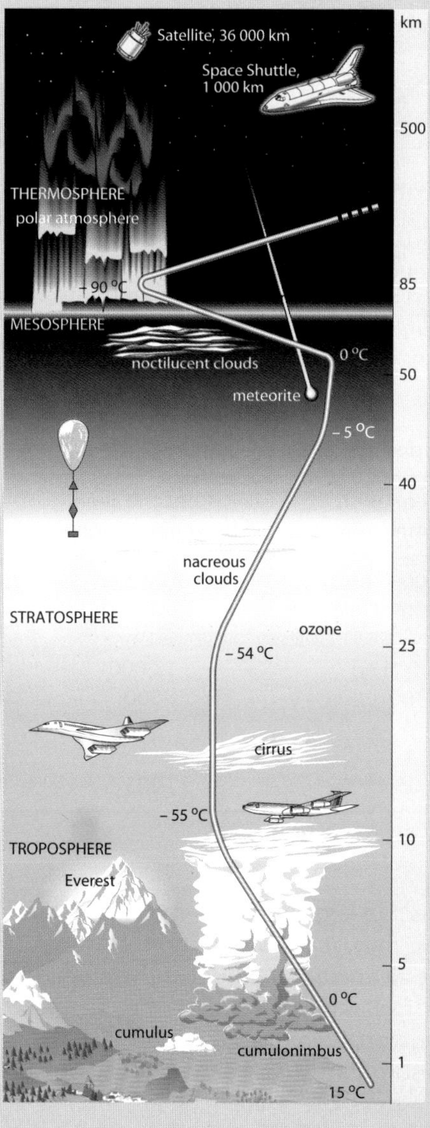

Figure 5. Vertical profile of Earth's atmosphere. Its structure, characterised by several maxima and minima of temperature, is partly due to the absorption of the Sun's ultraviolet radiation by ozone in the atmosphere. (From R. Chaboud, *Pleuvra, Pleuvra Pas*, Gallimard Jeunesse, 1994.)

As is the case with all planetary atmospheres, the pressure exerted by the Earth's atmosphere corresponds to the weight of a column of air on a surface per unit area, extending from the altitude of that surface (for example, at ground level where $z = 0$) to the upper limit of the atmosphere. It follows that, as altitude increases, and the volume of the column, its density and the force of gravity (g) all decrease, atmospheric pressure falls – according to a law approximating to the exponential $P(z) \sim e^{-z/h}$, where z is altitude and h the scale of height. In Earth's lower atmosphere, pressure falls by a factor e ($e = 2{\cdot}7$) at a height of about 8 km. On Mars and Venus, the corresponding altitudes are of the order of 10 km and 14 km. The temperature of the Earth's atmosphere falls with increasing altitude until an altitude of 12 km is reached. This is the *tropopause*. Within this region (the troposphere), zones nearer the surface being on average warmer than those at higher altitudes, heat is transferred upwards by convection. Warm air rises at the equator, where insolation is at its greatest, and descends at the tropics: the so-called Hadley circulation, a phenomenon also observed on other terrestrial planets, Mars and Venus (Figures 5 and 6).

Although water vapour is but a minor component of the Earth's air, representing less than a few per cent by volume, its role within atmospheric circulation is most important, because H_2O is the only

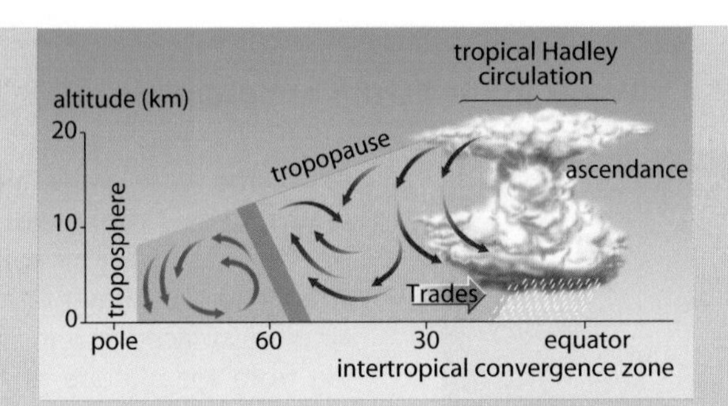

Figure 6. Atmospheric circulation. At the equator, warmed air rises, then cools, and clouds form. The dried air redescends, at about latitude 30°, to form the Hadley cell. Two other cells exist between latitudes 30° and 60°, and between 60° and the poles. The Coriolis effect, linked with the Earth's rotation, gives rise to the tropical trade winds. (From S. Jousseaume, *Climat d'Hier à Demain*, CNRS Editions-CEA, 1993.)

Table 1. The mean chemical composition of Earth's atmosphere.

Gas	Concentration by volume (ppmv)	Concentration by volume (ppmv)
Nitrogen (N_2)	781,000 (78%)	755,000
Oxygen (O_2)	209,500 (21%)	231,500
Argon (Ar), neon (Ne) and krypton (Kr)	9,400 (1%)	13,000
Carbon dioxide (CO_2)	300	500
Helium (He)	5	1
Methane (CH_4)	1	1
Hydrogen (H_2)	05	005

ppm = parts per million

molecule capable of condensation, present in large amounts in all three states: liquid, solid and vapour. As equatorial air rises, it cools, causing the formation of clouds and eventual precipitation. The air cooled at altitude is therefore drier as it descends over the tropics, which explains the relatively dry nature of tropical zones when compared with equatorial regions.

At the surface of the oceans, the absorption of solar energy leads to the evaporation of a certain quantity of water. The water content of the air increases until the mechanisms of condensation and precipitation occur, releasing into the atmosphere a quantity of energy equal to that used for evaporation. So the effect of the water cycle on Earth is to cool the surface and warm the atmosphere, attenuating the difference in temperature between the highest and lowest layers of the atmosphere.

Condensation of water vapour in the troposphere gives rise to the formation of the various types of cloud. At a low altitude (below 2 km) are stratocumulus types, at middle altitudes (1–6 km) are cumulus (cumulonimbus and altocumulus), characterised by their flat bases and columnar or 'cotton-wool' appearance. Cirrus, cirrostratus and cirrocumulus, typically filamentous in form, are found high in the atmosphere, between altitudes of 6 and 12 km. Observing clouds can provide an indication of the evolution of weather locally. For example, a build-up of cumulus may presage an storm, whilst the presence of cirrus – a sign of strong air currents at altitude – is often indicative of an impending change in the weather.

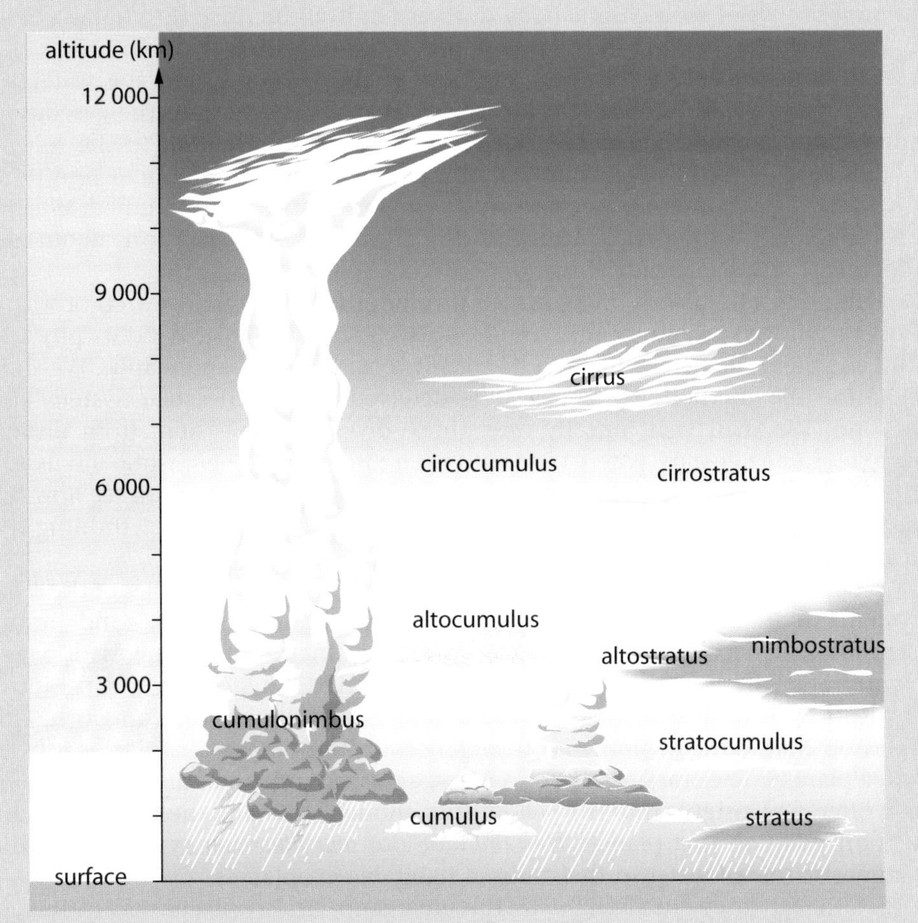

Figure 7. Condensation of water vapour in the troposphere creates the various types of clouds. At low altitudes, below 2 km, are found stratocumulus, and at mid-altitudes (1–6 km), cumulus (cumulonimbus and altocumulus), with typical flat bases and columnar or 'cotton-wool' forms. Cirrus, cirrostratus and cirrocumulus, typically filamentous in structure, occur in the upper troposphere at altitudes between 6 and 12 km.

... AND ELSEWHERE IN THE UNIVERSE

Given the evidence that water occurs throughout the Universe, we might expect that the chances of finding life out there are good. However, two major reservations immediately suggest themselves. First, we still do not know just how life started on Earth. The presence of liquid water was undoubtedly a determining factor, but not the cause in itself. In fact, there is little we can say about the probability of the emergence of life in a water environment. Also, the presence of water on a cosmic body does not of itself increase the chances of its harbouring life; the important thing is that it has to be in liquid form. Now, this seems to be an extremely rare condition in the Universe, and the water we observe on a great number of other celestial bodies is always in the gaseous or solid states and not liquid: the Earth seems to be an exception to the rule. A body's temperature and pressure determine the kind of water found. Conditions favouring liquid water are not directly observable on the bodies studied. We are, therefore, still a long way from knowing the answers to questions about the emergence of extraterrestrial life.

This book will take us on a journey through the Universe, in search of water. First of all, we shall study this commonplace little molecule, H_2O: its physical and chemical characteristics, and its cosmic formation and abundance. We shall examine the methods by which it is detected within the solar system and beyond. We shall visit, one by one, those diverse cosmic sites, from distant galaxies to nearer stars, where water has been discovered. Then, while discussing the overall mechanism of the formation of the solar system, we shall see how the H_2O molecule played a major part there too, with the ice line determining the natures of the terrestrial and the giant planets.

We shall also note the omnipresence of water in the various bodies of the solar system: in the giants with their rings and satellites, in comets, and in the terrestrial planets. Tracing the history of water in the atmospheres of Mars, Venus and the Earth, we will see how small differences in temperature, instigating the different states of water (vapour on Venus, liquid on Earth, and solid on Mars), led to immense divergences in the observed evolution of the three planets. The story of water on Mars, an aspect much discussed nowadays, can offer new insights into the possibility – though still but a theory – that there was once life on that planet. We conclude our *tour d'horizon* by looking at the diagnostic role of water in the study of habitable exoplanets.

So, water: an essential factor in the emergence and continuation of life on Earth, and a vital step in our quest for signs of extraterrestrial life.

1
A very simple molecule

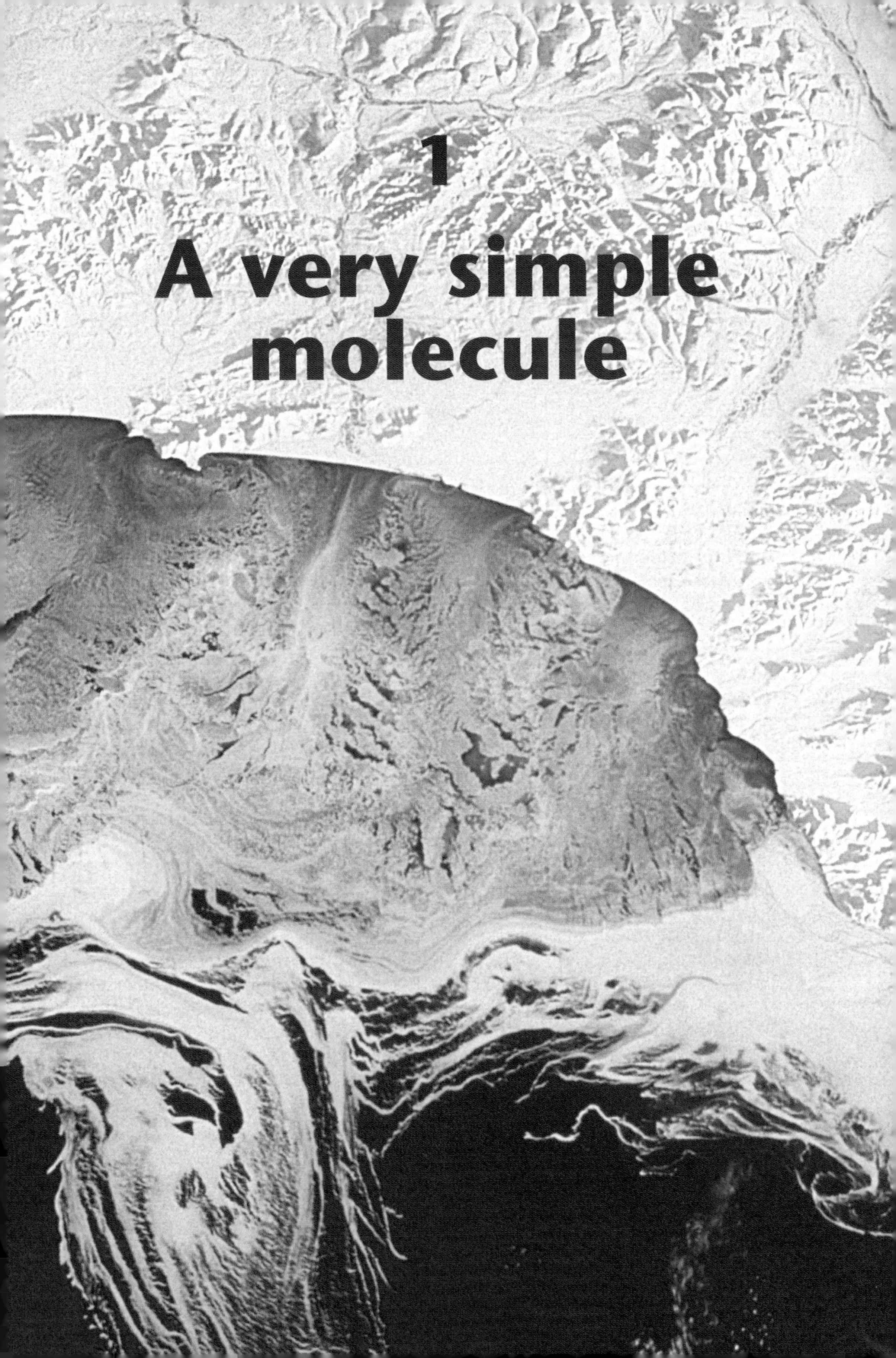

Two atoms of hydrogen, and one of oxygen. A simple little molecule, with physical and chemical properties precious for the temperate climate which fosters life on Earth.

How do we detect the presence of water elsewhere in the Universe? Spectroscopic analysis of heavenly bodies allows us not only to detect the presence of water in whatever state, but also to determine the temperature at which it formed – a very useful clue to the history of the celestial body in question.

THE H_2O MOLECULE

All that is needed to form a water molecule are two atoms of hydrogen and one of oxygen. Oxygen (O) – with its two possible chemical bonds (valencies) – forms the stable and electrically neutral H_2O molecule by combining with two hydrogen (H) atoms, each of single valency. In the conditions prevailing within the Earth's atmosphere, this apparently simple combination is obtained via three chemical reactions with a remarkably high energy budget. H and O atoms form, separately, H_2 and O_2 molecules; next, two H_2 molecules combine with one O_2 molecule to form two molecules of H_2O (Figure 1.1). This three-stage process of the construction of the water molecule from hydrogen and oxygen atoms is one of the most exothermic known, with an energy budget of –219 kcal per molecule. The formation of H_2O from two volumes of hydrogen and one of oxygen may occur spontaneously and explosively – thus the necessity to handle stocks of hydrogen carefully, given the natural presence of oxygen in the surrounding air. We shall see that within the Universe, and within the interstellar medium in particular, the formation of H_2O may involve other reactions.

Once the water molecule is formed, how stable will it be? On Earth it may be destroyed by solar ultraviolet radiation, which is very energetic. First, it breaks the bonds between the atoms. Higher-energy solar radiation in the far ultraviolet (Figure 1.7) can ionise the molecule by stripping away an electron, or ionising the atoms themselves. Generally, the stability of a molecule in the Earth's atmosphere can be measured according to its lifetime relative to this process of photodissociation (the time taken for a given number of molecules to be halved as a result of their destruction by solar radiation). In the case of water, this is about 36 hours. The water molecule is more stable than ammonia (NH_3), formaldehyde (H_2CO) and methanol (CH_3OH), but is less stable than carbon monoxide (CO), carbon dioxide (CO_2) and methane (CH_4). So, from the point of view of stability, the water molecule is nothing special. The same can be said of its thermal stability: it can resist temperatures up to 3,000°C before destruction.

Satellite view of the Kamchatka peninsula, eastern Siberia, Russia.

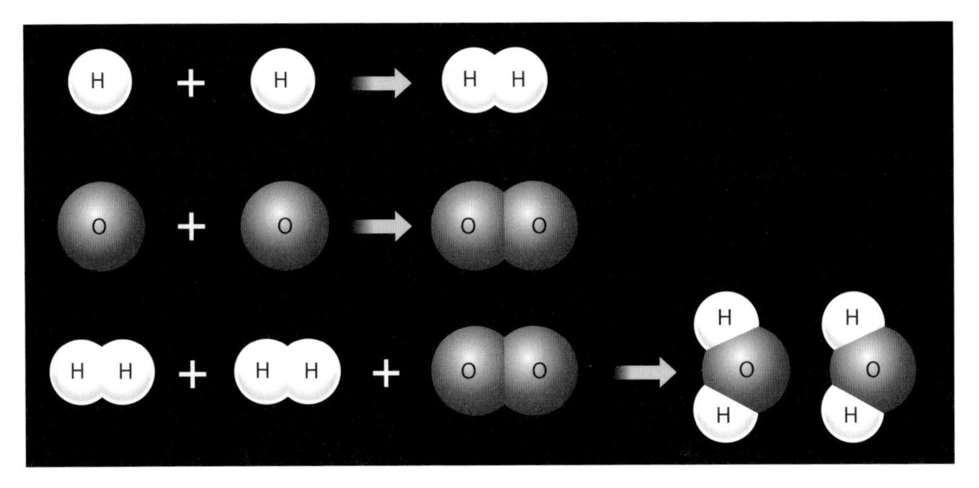

Figure 1.1. In the conditions prevailing in Earth's atmosphere, the water molecule arises from a chain of three chemical reactions with a remarkably high energy budget (–219 kcal per molecule). From atoms of H and O, separate H_2 and O_2 molecules are formed. Two H_2 molecules then combine with one O_2 molecule to form two H_2O molecules.

THE VARIOUS STATES OF WATER

Water is the only component of the atmosphere found naturally, in equilibrium, in all three states: gas, liquid and solid. As far as we know, this is unique to our planet. The oceans hold almost all our planet's stock of water, while its climate, and the progress and continuation of life on Earth, are determined largely by continuous exchanges of water between oceans, lakes, rivers, polar ice sheets and the atmosphere (the water cycle). Why should this be? The phase diagram (Figure 1.2) offers the answer.

The curves on the diagram indicate the boundaries between the different states of water, as a function of temperature and pressure. They lie between the domains where pressure and temperature cause water to be in solid, liquid or gaseous form. The triple point, at the intersection of the three curves, corresponds to a temperature of 0.01°C (273.16 K) and a pressure of 0·006 atmospheres. Mean atmospheric temperature is quite close to that of the triple point, allowing the coexistence of the three states. However, since it is slightly higher, most of the Earth's water is in the liquid state; though the other states are not disallowed, mainly because of seasonal temperature fluctuations. So, for example, we see polar ice sheets growing every winter, only to shrink again as spring comes.

Ice

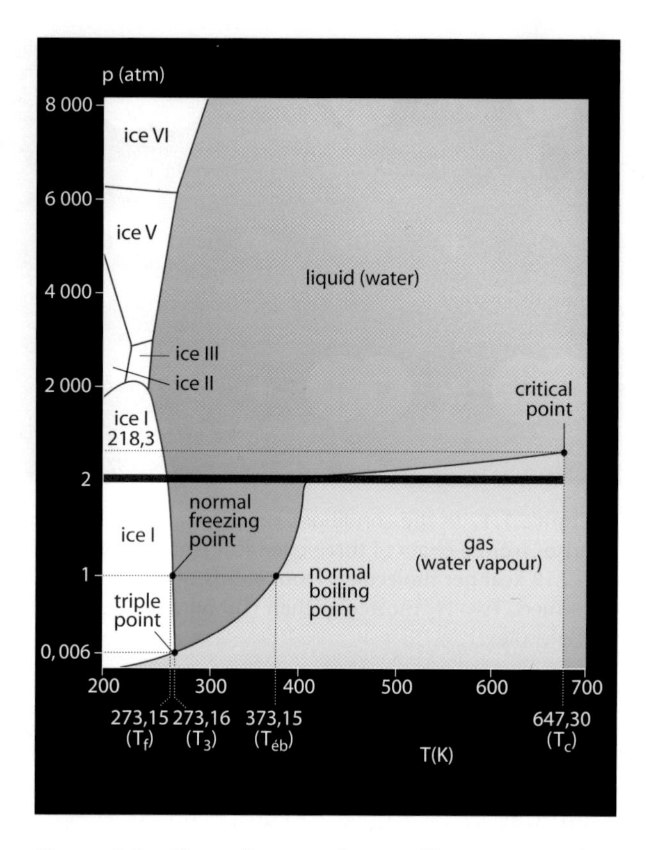

Figure 1.2. Phase diagram of water. The curves on the diagram indicate the boundaries between the different states of water, as a function of temperature and pressure. The Roman numerals represent different types of crystalline ice. The scale of the vertical axis changes at 2 atmospheres.

In its solid form, water as ice may assume many different guises. It may be amorphous – when its atoms are distributed in a random way – or crystalline – exhibiting various structural forms according to local temperature and pressure. The crystalline form is what we observe on Earth and on other planets of the solar system. The amorphous form is more often seen in the interstellar medium. This differentiation is not, *a priori*, surprising. We might expect the amorphous form to be present in the cold, rarefied interstellar medium, since a certain amount of energy (at least comparable to that found on Earth) is required to organise molecular systems into a crystalline structure. It should be noted that water ice can be found in another kind of arrangement: *clathrates*. Solid matrices of water molecules are able to entrap another element; for example, a molecule of methane or carbon dioxide, or an atom of argon. Clathrates of methane and carbon dioxide exist on Earth: in the case of methane, a molecule of CH_4 is trapped within a matrix composed of about six water molecules. Clathrates may have played an important part in the formation of the planets and satellites of the outer solar system, and of comets (see Chapters 3 and 4).

The remarkable properties of liquid water

In its liquid form, water has a specific and very important property: it is a very effective solvent. Thanks to its high dipole moment μ (Figure 1.4a), liquid water is able to dissociate molecules electrically into positive or negative ions (anions and cations). This remarkable property has had an effect essential to the

The different forms of water ice

On Earth, water ice occurs naturally in the form of a crystalline hexagonal network. Oxygen atoms occupy the joints in a network of hexagons, which when superimposed to reach an observable scale become familiar snowflakes. Hexagonal ice forms at temperatures between 200 and 273 K (–73.15°–0.15° C). Other kinds of ice, known as allotropes, exist at lower temperatures, and have been studied under laboratory conditions. Between 135 and 200 K, ice takes on a cubic form. Below 135 K, it is amorphous, the molecules trapped within showing no regular structure. This last type is observed within the interstellar medium.

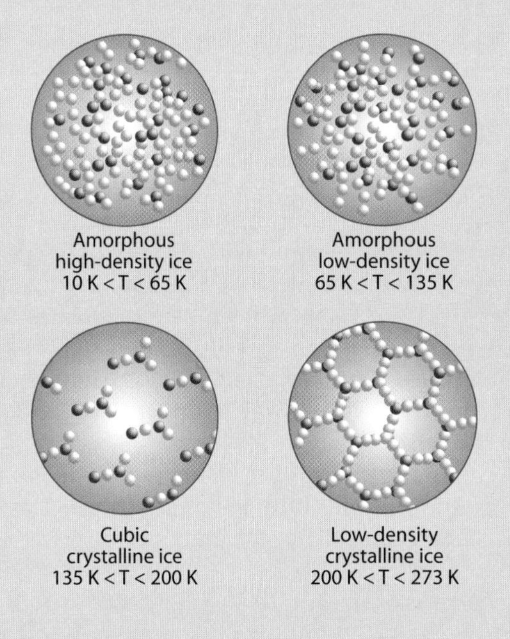

Figure 1.3. The various forms of water ice. The amorphous variety, found at temperatures below 135 K, is observed in the interstellar medium.

In crystalline ice, organic substances are driven out, but studies of amorphous ice have shown that they may be able to react within it – a factor encouraging the development of interstellar chemistry. It is possible that some of these substances are carried on comets, which can retain a fraction of their amorphous ice in spite of warming as they near the Sun, and that they have been delivered into the Earth's atmosphere; but this remains only a theory.

evolution of our planet and the life upon it: it has meant that great quantities of carbon dioxide present in the primitive atmosphere of the Earth have been dissolved into its oceans, there to react with the calcium oxide (CaO) of the rocks to form carbonates such as calcium carbonate ($CaCO_3$). This regulation of the abundance of carbon dioxide has in its turn led to the stabilisation of the temperature of the atmosphere throughout the lifetime of this planet.

Another special feature of liquid water is its low viscosity, which makes it a highly mobile liquid, capable of soaking into the soil. This is a most important

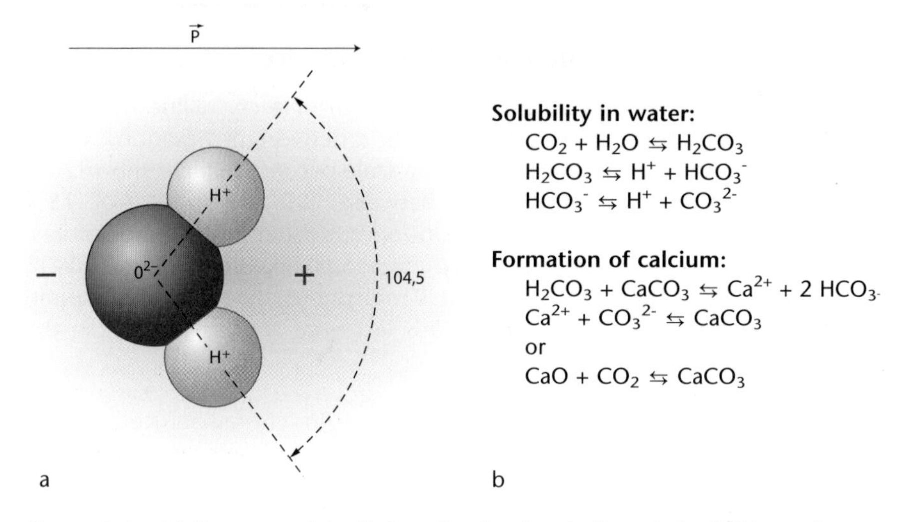

Figure 1.4. (a) Structure of the H_2O molecule, electrically polarised; (b) reactions leading to $CaCO_3$, formed from carbon dioxide dissolved in the oceans.

property, vital to the survival of living things. Water – with its very high specific heat and low thermal conductivity – has considerable thermal inertia. This brings great climatic consequences, and the Earth's oceans are regulators of its temperature. Also, the latent heats of vapourisation and fusion of water are high compared to those of other molecules. The result of this is that, in our atmosphere, temperature differences in the lower troposphere and the tropopause are markedly reduced.

Finally, the remarkable thing about the density of water is that it is at its maximum at 4° C, so that ice floats upon liquid water. Changes in state, due to changes in insolation, occur at the surface, and the oceans do not freeze from the bottom upwards. This property ensures that species living at the bottom of frozen rivers survive, and has also played a major part in the way in which the Earth's atmosphere has evolved.

GREAT COSMIC ABUNDANCE

Water is abundant on Earth. But is this the case throughout the Universe? To answer this question, we have first of all to consider the abundances of its component elements: hydrogen and oxygen.

The astronomers' study of stars and galaxies at different stages of their evolution has led them to an appreciation of the main chapters in the story of the Universe, within the Big Bang model. All the cosmic phenomena that we observe today are the result of an initial explosion which happened between 12 and 15 billion years ago. At first, temperatures reached unimaginable levels (of the order of 10^{40} K), and the Universe consisted only of elementary particles with

The Big Bang

The standard cosmological model known as the Big Bang, derived from models by Friedmann and Lemaître, describes the first stages of an homogeneous, isotropic Universe in expansion, culminating in the formation of the galaxies. In its broad lines, the Big Bang theory is now widely accepted by the scientific community. Observational evidence comes partly from the recession of the galaxies – which, as Edwin Hubble revealed, move away faster as their distance increases – and partly from the discovery of the cosmic background radiation at 2.7 K, the echo of the primordial explosion which gave birth to the Universe about 15 billion years ago. First identified by American astronomers Arno Penzias and Robert Wilson in 1965, the cosmic background radiation was measured very accurately in 1992 by the COBE (Cosmic Background Explorer) satellite, which confirmed its remarkably isotropic character.

Taking t_0 to be the instant of the initial explosion, we can only begin to describe the Universe after an elapsed time of 10^{-43} s. It is impossible to describe phenomena before this 'Planck time', determined by the uncertainty principle of quantum mechanics. At $t = 10^{-43}$ s, the temperature is 10^{32} K. Thereafter the Universe cools adiabatically (without gain or loss of heat). At first it consists of photons and a mixture of elementary particles (quarks, electrons, and so on) and their antiparticles. As the temperature decreases, so particle–antiparticle pairs mutually annihilate, which is possible because the thermal energy (proportional to temperature) is now less than the rest energy of the particles (uniquely as a function of their rest mass). At $t = 10^{-9}$ s ($T = 10^{13}$ K) the quarks join together in threes to form protons and neutrons. At $t = 10^{-9}$ s, the temperature is 10^9 K. Electrons and their antiparticles, positrons, annihilate each other, emitting very energetic gamma-rays. A series of nuclear reactions leads to the formation of light elements (helium, deuterium, lithium), the primordial nucleosynthesis representing the creation of matter *en masse*: 75% hydrogen nuclei, 23% helium nuclei, with traces of the others. The predominance of hydrogen seen in today's Universe is nothing but the consequence of this early nucleosynthesis.

The temperature was then high enough for protons and electrons to associate permanently, but it continued to fall and, 300,000 years later, stood at 3,000 K. Electrons were then able to combine with nuclei to form the first hydrogen and helium atoms. Now interacting less with matter, photons were free to propagate – an epoch known as the 'decoupling' of matter and radiation. It is probable that the first galaxies appeared during the first or second billion years after the Big Bang; but how this came about is still a matter of debate.

extremely high energy. Then, a cooling process began – at first rapidly and then more and more slowly, as the Universe expanded. The first atomic nuclei (hydrogen and helium) formed, and later, galaxies and the earliest stars. Within the stars, heavier elements were made.

In its early stages, a star burns its hydrogen as a result of nuclear reactions unleashed when the temperature at its core reaches values of about 10 million degrees. When the hydrogen has been transmuted into helium, nucleosynthesis continues with the formation of carbon, nitrogen and oxygen. Thereafter, if the core temperature continues to rise, other, heavier elements are synthesised, until iron is formed. Only the most massive stars can create even heavier elements, which are ejected into space in supernova explosions, ending the lives of these stars.

With the exception of the lightest atoms – essentially, hydrogen and helium – formed by primordial nucleosynthesis as direct products of the Big Bang, all the elements in the Universe, including the atoms of our own bodies, were synthesised within stars, and ejected by them into the cosmos as they died. These elements were later integrated into other, nascent stars. This immense mixing process has led to all the atoms we know of – atoms whose relative cosmic abundances we are studying by means of astronomical observations.

As a general rule, the lightest atoms are the most abundant, because their formation required less energy (the temperatures were lower within the stars). Pride of place is assumed by hydrogen: 75% of the total mass of the Universe is hydrogen, followed by helium (about 23%). All the other elements make up the remaining 2%. Among them, the lighter elements such as carbon, nitrogen and oxygen are favoured. Cosmic abundances of oxygen and carbon are globally comparable, while nitrogen is less abundant (Figure 1.5).

Here, therefore, is a factor favouring the formation of water in great quantities: the great preponderance of hydrogen, with oxygen – the most abundant heavier element able to associate with it. Also, lesser though still important quantities of ammonia (NH_3) and methane (CH_4) are the result of the association of nitrogen and carbon with hydrogen. Thus arises another quality of water ice, as distinct from the ices of ammonia and methane. As temperature falls, water is the first to solidify into ice, while ammonia, methane and other molecules freeze at lower temperatures. We shall see that, because of this property, water played an essential role in the formation of the solar system, and in the condensation of the planetesimals from which planets formed.

How does water form within the interstellar medium?

How is the water molecule formed within the interstellar medium? Since hydrogen is the most abundant element, the most common reactions will be those involving H and H_2. However, the temperature here is low (less than about 100 K), implying that reactions requiring no energy input (mostly exothermic reactions) are the most likely to occur. In the diffuse and tenuous interstellar medium, reactions between hydrogen and oxygen in the gaseous state depend upon the formation of the very reactive ion H_3^+. This ion – at the heart of chemical reactions within the interstellar medium – is obtained through ionisation of H and H_2 by cosmic rays. It reacts very readily with atomic oxygen to form OH^+ and H_2, and thence H_2O^+ and H. The H_2O^+ ion in turn reacts with H_2, forming the H_3O^+ ion and H. Finally, H_3O^+ reacts with free electrons to form the OH radical and the neutral H_2O molecule.

As well as this diffuse matter, interstellar space also contains the cool, dense molecular clouds where stars form. Here, the temperature, of the order of 100 K in the diffuse medium, does not exceed a few tens of K. Water can also form directly as ice, via chemical reactions on the surfaces of the interstellar dust grains present. These grains are made of carbon and silicon, and to them adhere the most abundant atoms and molecules, coating them in ice. The great mobility of hydrogen atoms upon cold surfaces encourages chemical reactions and the formation of the H_2O, CH_4 and NH_3 molecules.

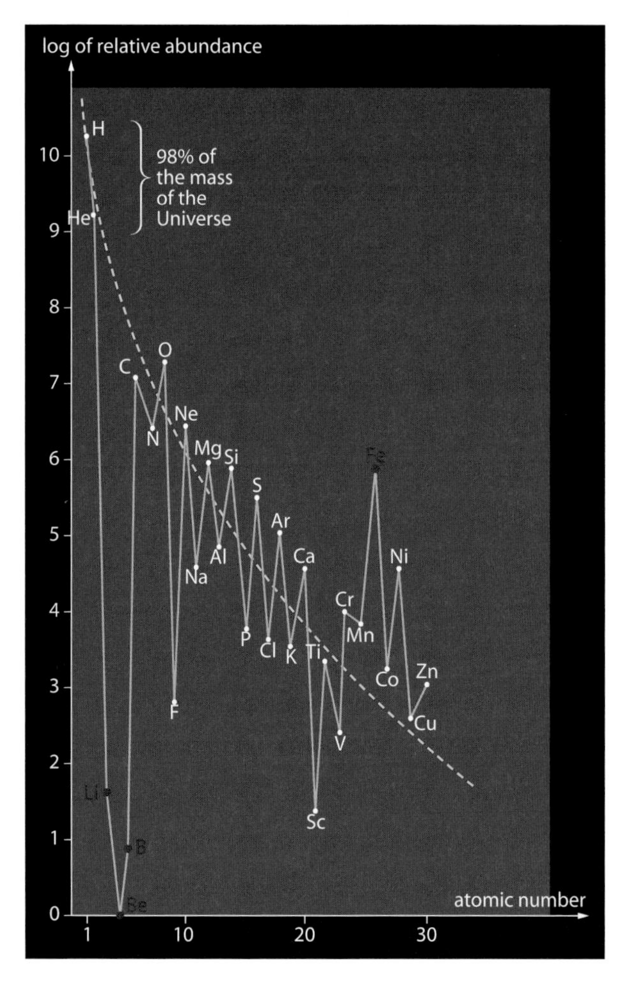

Figure 1.5. Curve representing the abundance of chemical elements in the Universe. Only light elements, up to and including zinc, are shown. These abundances are measured principally in the Sun, with the exception of deuterium, in the case of relatively light elements (C, O, N), and in meteorites in the case of heavier elements. Li, B and Be are very rare, while Fe (in red) is very abundant. (From C. Allègre (modified), 1985.)

Why look at the spectra of stars?

Thanks to spectroscopy within the laboratory, we can determine the spectral signature of a given substance, whether atomic or molecular. Analysis of a star's

Spectroscopic analysis

A ray of white light passing through a prism is split into several rays, revealing the colours of the rainbow. Each colour corresponds to an electromagnetic radiation of frequency ν and wavelength λ (= c/ν) between 0.4 μm (violet) and 0.7 μm (red). This range of frequencies is but a small part of the electromagnetic spectrum, whose wavelengths may be as short as a 0.1 Å and as long as hundreds of kilometres. We know, too, that all electromagnetic radiation (frequency = ν) consists of particles (photons) of energy $h\nu$ proportional to the frequency (h being Planck's constant). Due to the discrete structure of their energy levels $E_{i'}$, atoms and molecules can emit or absorb radiation only at frequencies $\nu_{ij} = | Ei - Ej | /h$. Therefore, the spectrum, representing the energy structure of each atom or molecule, allows us to distinguish one chemical compound from another.

Figure 1.6. White light, split by a prism, forms a continuous spectrum comprising all visible frequencies.

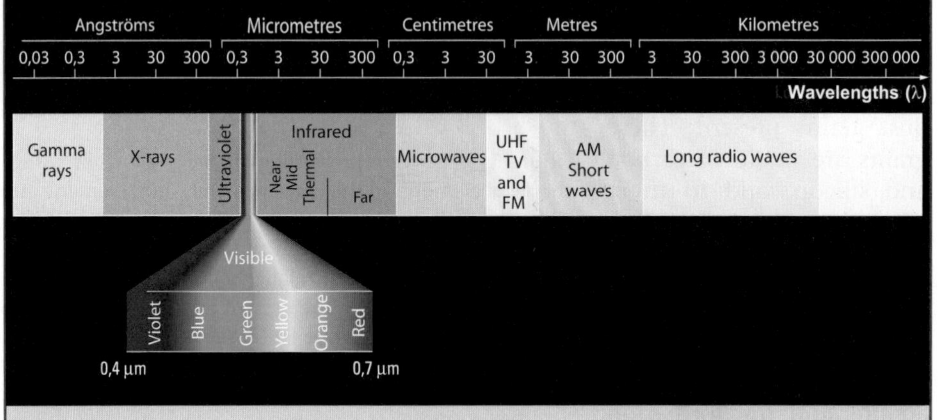

Figure 1.7. Electromagnetic radiation.

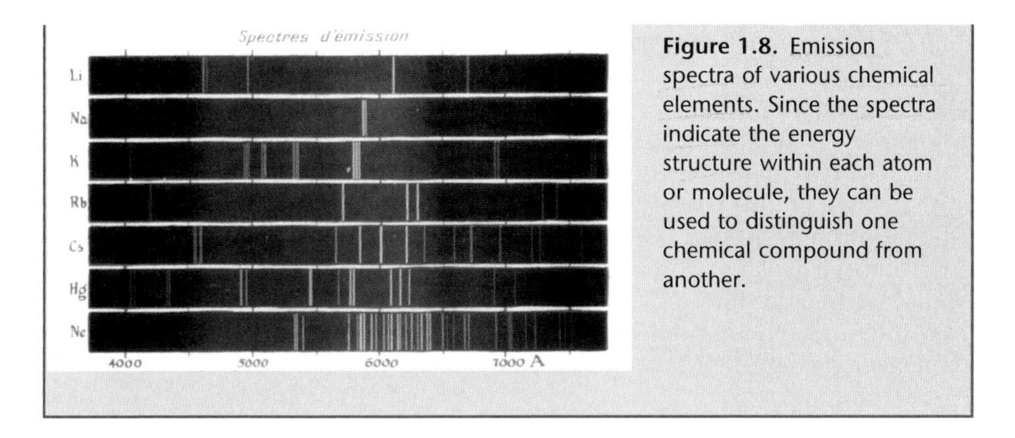

Figure 1.8. Emission spectra of various chemical elements. Since the spectra indicate the energy structure within each atom or molecule, they can be used to distinguish one chemical compound from another.

spectrum reveals the presence of a given compound. In reality, the observed stellar spectrum may differ from similar laboratory spectra because of the physical conditions and chemical composition of the milieu traversed by the radiation from the celestial body (Figure 1.9). It is sometimes difficult to reproduce the exact conditions of the observation of that body and to determine, with spectroscopy, the abundance of a certain constituent, but its identification is normally unambiguous.

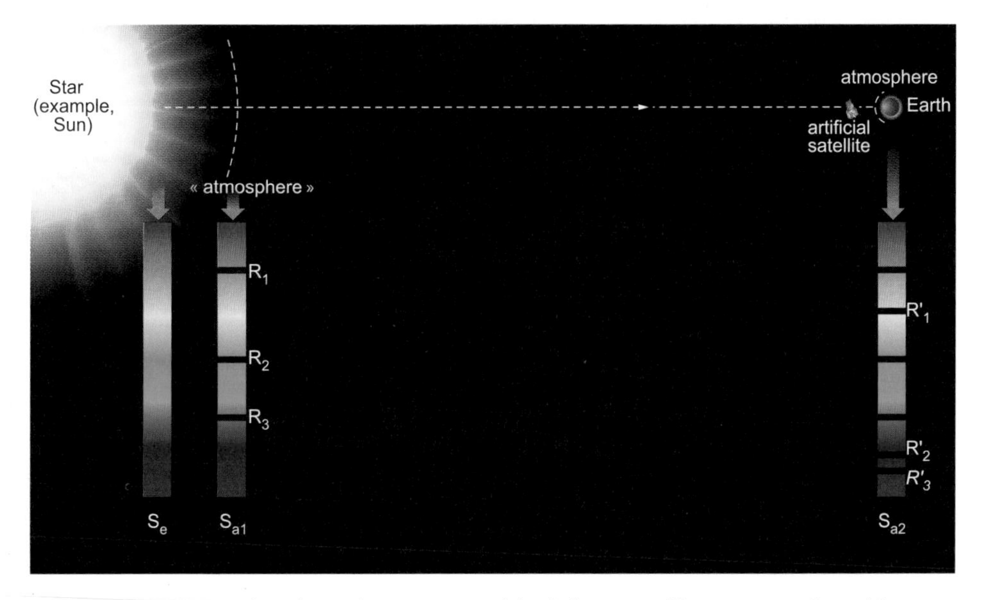

Figure 1.9. Spectral analysis. The spectrum of the light emitted by a star or reflected from a planet (S_e) is continuous. The absorption spectra of stellar or planetary atmospheres (S_{a1}, S_{a2}) exhibit dark lines superimposed on the original spectra. These lines are the signatures of the chemical composition of the atmospheres through which the radiation has passed, and which may be inferred from comparison with spectra established on Earth. S_e, emission spectrum; S_a, absorption spectrum. (Based on P. Miné, 2001.)

THE SPECTRUM OF THE WATER MOLECULE

Returning to the water molecule H_2O: its structure – well known in the laboratory – is not linear, but in the form of a triangle. At the apex is the oxygen atom, with two hydrogen atoms at a distance of 95.7 pm and forming an angle of 104°.5. If we assume that each atom of the molecule moves around its point of equilibrium, we see that the distances within the configuration can change. A simple description of the vibrations of the molecule can be derived from appropriate combinations of atomic modes of vibration. *Normal modes* correspond to synchronous but independent vibrational movements of atoms; they may be excited without the excitation of another normal mode. In the case of the H_2O molecule, as with most molecules, the normal-mode frequencies involved lie within the infrared. Two of the three modes associated with water (Figure 1.10) correspond to a wavelength of about 2.7 μm (elongation modes), while the third corresponds to a wavelength of about 6.2 μm (angular deformation mode).

The infrared spectrum of water is therefore characterised by lines around these three wavelengths.

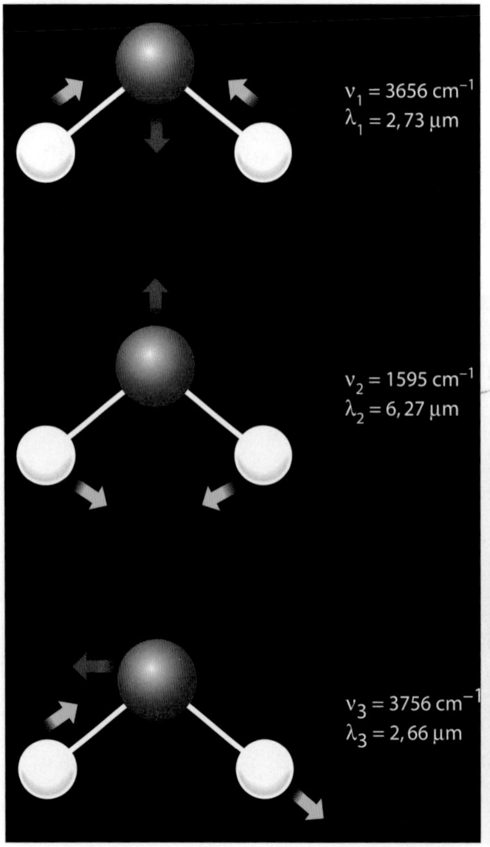

Figure 1.10. The three normal vibrational modes of the H_2O molecule. The ν_2 mode is principally a deformation mode and is produced at a wave number (the inverse of the wavelength) lower than that of the two others, which are principally elongation modes.

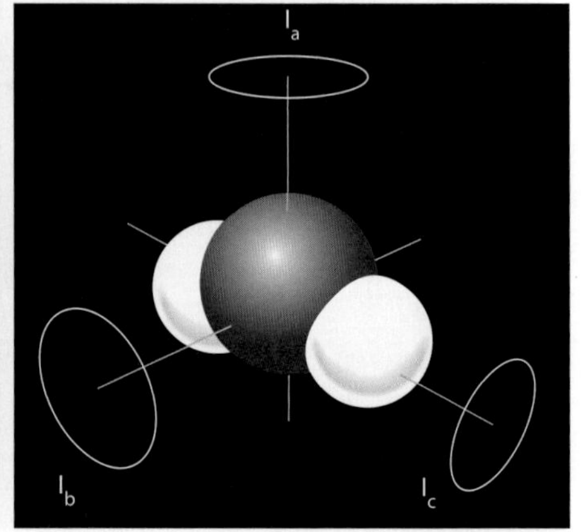

Figure 1.11. The three normal axes of rotation of the water molecule, an asymmetrical rotator. The three axes meet at the barycentre of the molecule.

There are other lines, too, beyond 20 μm in the far-infrared, and at microwave and radio wavelengths. This is due to the three rotational movements within the water molecule (Figure 1.11). So, whenever we use a microwave oven, we are employing microwave radiation to stimulate the water molecules in the food to higher levels of rotational energy!

A closer examination of the lines of the vibrational spectrum of water in the near and middle infrared ($1 < \lambda < 20$ μm) reveals that they can be subdivided into a great number of very closely spaced lines. For this reason, molecular spectra are often known as *band spectra*. These are the result of *rotational transitions* accompanying each *vibrational transition*. A vibrational transition involves a sudden modification of the length of the bond, causing an acceleration or deceleration of molecular rotation. The same principle applies to an ice skater, spinning more rapidly with arms held close to the body, and more slowly if they are held out. The spectrum in question is called a *vibration–rotation* spectrum.

Studying the vibrational and rotational spectra of stars and planets can therefore provide us with clues to the possible presence of water there; and the same is true of most of the other gaseous substances having specific spectral signatures, such as H_2, CH_4, NH_3, CO, CO_2, and so on.

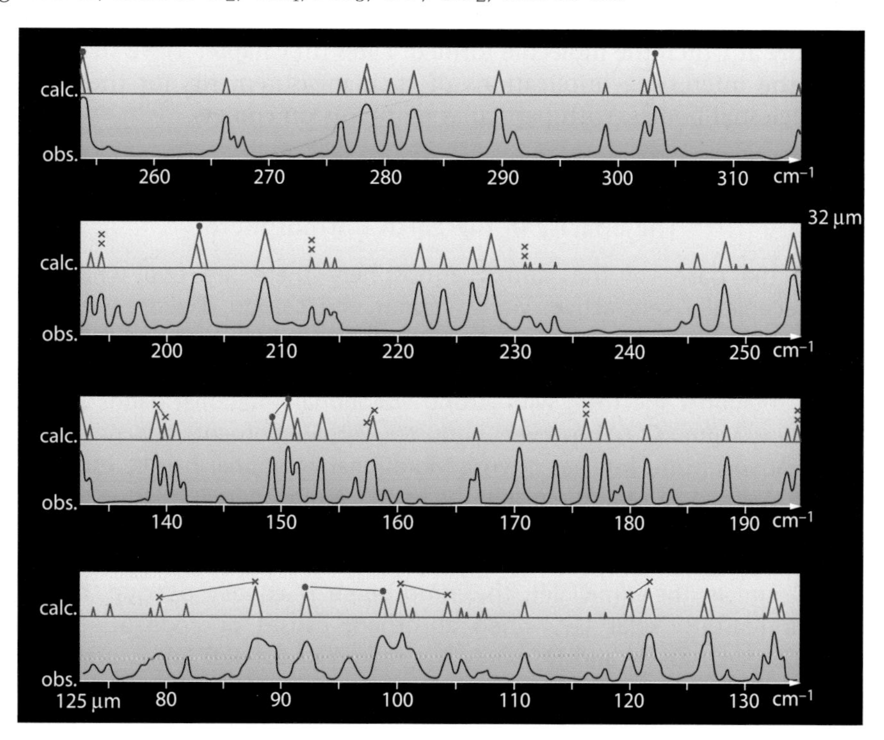

Figure 1.12. The rotational spectrum of water. The laboratory spectrum is compared with the theoretical positions of transitions between 310 and 130 cm^{-1} (between 32 and 77 μm). These are the spectral signatures used to identify H_2O in the infrared spectra of celestial bodies. (From G. Herzberg, *Infrared and Raman Spectra*, Van Nostrand, 1945.)

THE *ORTHO* AND *PARA* STATES OF WATER

The H_2O molecule has yet more special characteristics; for example, its *ortho* and *para* states. These terms stem from two special states of dihydrogen (H_2) – a molecule composed of two hydrogen atoms, each with one proton (or nucleus) and one electron orbiting it. The rotational direction of the proton is defined by its spin value, +1/2 or −1/2. If the nuclear spins are in opposite directions, the H_2 molecule is known as para; and if the vector sum of the spins is non-zero, it is ortho. The same definitions apply to the H_2O molecule according to whether or not the nuclear spins of the two H atoms are in opposite directions.

This distinction is very interesting. Through it we can show that there are three times as many ways of attaining the ortho state than the para. This implies that the intensities of the spectral lines corresponding to rotational and vibrational–rotational transitions of the two types of H_2O, occurring at slightly different wavelengths, are in the ratio 3:1. This allows us to measure, spectroscopically, the relative quantities of each of the two types. The relationship between the abundances of the two types depends upon the temperature at which the molecule forms – and no mechanism can later modify it. Measuring the ortho:para ratio of water therefore provides a direct indication of the temperature at which the molecule formed. Later (in Chapter 4) we shall further examine the interesting implications of such measurements for the study of water in celestial bodies, with particular emphasis on comets.

The opacity of the Earth's atmosphere

Among the gases of the earth's atmosphere there are two which are spectroscopically very active: water vapour and carbon dioxide, while the two principal gases, N_2 and O_2, are practically inert. Both H_2O and CO_2 have very rich spectra, and H_2O, with its strong dipole moment, combines a very intense rotational spectrum with strong vibrational–rotational bands around 2.7 and 6.2 μm. CO_2 however, with no dipole moment, has no purely rotational spectrum, but has intense vibrational–rotational bands, especially around 15 and 4.25 μm. As a result, the Earth's atmosphere is opaque to most infrared radiation from outside, with the exception of a few infrared windows, traditionally labelled J to N, around 1.25 μm, 1.65 μm, 2.2 μm, 3.5 μm, 4.5 μm and 10 μm. In the ultraviolet, the atmosphere is equally opaque, but for different reasons: solar UV, when absorbed, causes photodissociation of molecules such as O_2 and O_3. Only the region of 0.4–0.9 μm is practically transparent, causing the warming of the Earth's surface and constituting almost the total solar energy received. It is possible to observe the domains of the near and mid-infrared from the ground due to the atmospheric windows, but not the far or submillimetre domains, so astronomers have recourse to measurement in space or from aircraft. During the 1980s and 1990s, NASA's flying observation platforms (on a Lear Jet and the Kuiper Airborne Observatory) carried out high-altitude observations, the KAO reaching a height of 14 km. However, the best data have been obtained from Earth orbit – notably by the ISO satellite. Since then, other satellites (Submillimeter Wavelength Astronomical Satellite, and Odin) have concentrated on the particularly intense submillimetre water transition at 557 GHz (λ = 538 μm). The European Herschel satellite, to be launched in 2007, will take up the baton from ISO with a high-resolution all-sky survey in the submillimetre.

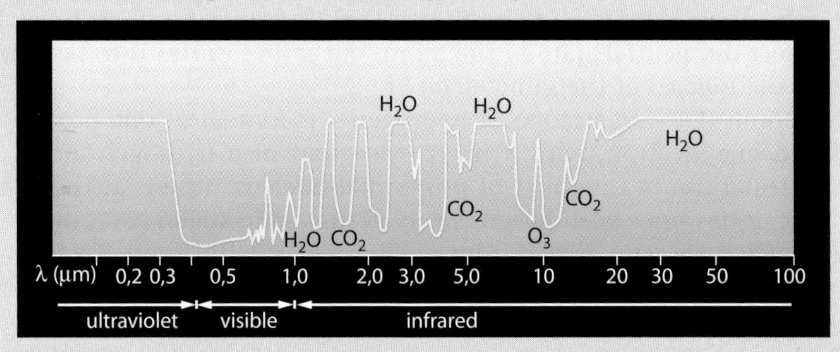

Figure 1.13. Absorption in the Earth's atmosphere, from ultraviolet to infrared. When absorption equals 1, the atmosphere is completely opaque, as is the case for the ultraviolet and far-infrared domains. If it is near zero, the atmosphere is transparent; for example, in the visible. There are several atmospheric windows in the near and mid-infrared.

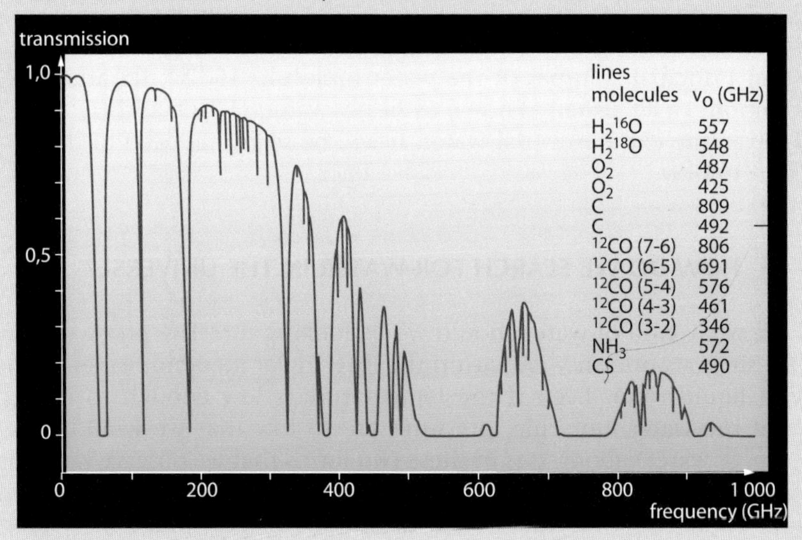

Figure 1.14. Transmission through the Earth's atmosphere in the radio, millimetre and submillimetre domains. Most atmospheric absorption is due to H_2O and its isotopes. The windows used by astronomers are found at 80–110 Ghz (around λ = 3 mm), 130–180 Ghz (λ = 2 mm), and 200–280 Ghz (λ = 1.3 mm). Observations in the submillimetre windows (610–690 Ghz, λ = 460 μm, and 800–890 Ghz, λ = 350 μm) require an excellent high-altitude site and very dry atmospheric conditions.

HEAVY WATER

One final characteristic of the H_2O molecule is also of great import to those studying water elsewhere in the Universe. We know that the nuclei of atoms consist of protons, with positive charge, and neutrons, identical to the protons but electrically neutral. The electrons orbiting these nuclei are of negligible mass compared with the protons, and have negative charge. Protons and electrons exist in equal numbers within the atom, rendering it electrically neutral. However, the nucleus may contain an excess of one or more neutrons, which does not effect the neutral state of the atom, but rather causes it to be heavier. Such atoms are *isotopes* of the same element.

So, deuterium (D) – an isotope of hydrogen – is formed of one proton, one neutron and one electron, with a mass twice that of a hydrogen atom. Like hydrogen, deuterium is a product of primordial nucleosynthesis at the time of the Big Bang. It has since been progressively destroyed in stellar cores by nuclear reactions which have transformed it into helium-3 (^3He, the figure 3 referring to the atomic mass), which in its turn is transformed into helium. Even though the cosmic abundance of deuterium is very small compared with that of hydrogen – the D:H ratio is without exception less than one in 1,000 – this isotope is a key element in our investigation of water in the Universe. We shall be looking at it later.

The two oxygen (^{16}O) isotopes (^{17}O and ^{18}O) are similarly far less abundant than ^{16}O. The ratios ^{16}O:^{18}O and ^{16}O:^{17}O are respectively of the order of 500:1 and 2,700:1. Combining the various isotopic forms of hydrogen and oxygen, we obtain four principal isotopes of the water molecule: $H_2{}^{16}O$, far and away the most common; HDO, usually known as heavy water; $H_2{}^{17}O$; and $H_2{}^{18}O$.

As we examine extraterrestrial water, it will be seen that each of these forms has its part to play.

HOW DO WE SEARCH FOR WATER IN THE UNIVERSE?

First of all, what kind of water should we be looking for? The phase diagram in Figure 1.2 suggests one answer: within the interstellar medium, water cannot be present in liquid form. Even if the temperature is low enough to ensure the stability of the water molecule, pressures are so low that we would expect to observe ice, or water vapour. It is in these two forms that we observe water on the planets, satellites, ring systems and comets of the solar system. Two notable exceptions come to mind, however. Observations suggest that liquid water ran across the surface of Mars early in its history, and there is strong evidence that an ocean of liquid water exists beneath the frozen surface of Europa (one of Jupiter's four Galilean satellites). (These cases will be revisited in Chapter 7, as they bring new perspectives to the search for extraterrestrial life.)

In summary, our search for cosmic water will principally involve its gaseous (water vapour) and solid (ice) phases.

The search for water vapour

The spectroscope is the tool used in the search for water vapour associated with celestial bodies. In their spectra we try to find the signatures of the transitions characteristic of the molecules investigated (see fact box, p. 24). These transitions may result in a maximum of light emitted at certain wavelengths (emission lines), or at a minimum (absorption lines). The physical conditions existing in the vicinity of the celestial body will determine the nature of the spectrum, whether it is of the emission or absorption type. In either case, however, the observation of the spectral signatures (exactly as in laboratory examples) provides unambiguous proof of the existence of the molecule in the object observed.

Rotational and vibrational transitions of water vapour are to be found, as we have seen, within the infrared and radio domains, as indeed is the case with most molecules. Other transitions may take place at higher frequencies, in the visible. These *electronic transitions* correspond to changes in energy levels in the electrons of molecules. Still more energetic radiations, in the ultraviolet, lead to the dissociation or ionisation of molecules.

It is, however, within the infrared, millimetric and radio domains that we look for water vapour, since it is here that the lines are at their most intense, and therefore easiest to observe.

One problem arises, however. Water vapour is present in Earth's atmosphere, in quantities varying with site and season. At sea level, amounts of atmospheric water vapour are sufficient to entirely absorb radiation at infrared and millimetric wavelengths, with the exception of some rare spectral windows. It is therefore no use trying to look into space at these wavelengths from the ground. This is the reason why the study of cosmic water vapour has progressed only recently, since the launching of Earth-orbiting infrared observatories. The major instrument of this research has been the European Infrared Space Observatory (ISO) satellite, which operated between 1995 and 1998. We owe to the ISO our knowledge that water is omnipresent in the Universe.

The discovery of the existence of water elsewhere in the Universe does not, however, date only from the achievements of the ISO. In spite of the obstacle presented by terrestrial atmospheric absorption, radio astronomy had been able to detect the presence of water vapour in various cosmic locations. Since the late 1960s, measurements have been possible within regions of star formation at a frequency of 22 GHz, corresponding to a transition in the hyperfrequencies (and therefore observable from Earth) characteristic of the H_2O molecule. In the intervening period, several transitions in the radio and millimetric domains, associated with various types of astrophysical target, have been observed from Earth.

The search for ice within the solar system

What is the status of ice in the Universe? Observing it spectroscopically is less problematical for astronomers, since its presence in the Earth's atmosphere, in

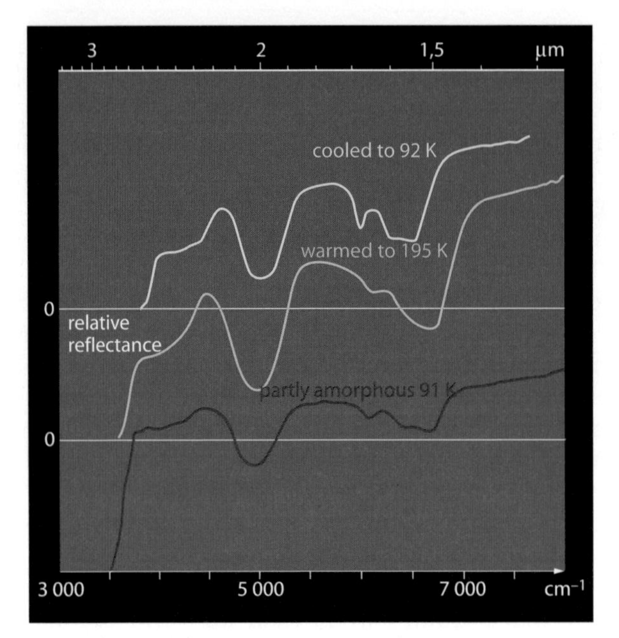

Figure 1.15. The spectrum of water ice in the near-infrared. This laboratory spectrum shows the characteristic spectral signatures of water ice at 1.5 and 2.0 μm. Another, more intense signature is also present at 3.1 μm.

cirrus clouds, tends not to affect the best astronomical observatory sites, usually found at high altitude in dry places. The spectroscopic signatures of ice in the Universe can therefore be studied from the Earth's surface. The water-ice spectrum displays characteristic lines in the near-infrared, particularly around a wavelength of 3 μm (Figure 1.15). Thanks to the ISO, other signatures at longer wavelengths, in the mid- and far-infrared, have been detected on certain bodies and within the interstellar medium.

The quest for evidence of water in the solar system is not new. For centuries, astronomers, physicists and philosophers have wondered about the possibility of life on neighbouring planets; and, of course, they were interested in the likelihood of water being there. They were too often disappointed: Mars and Venus, our two nearest neighbours, hold only tiny quantities of water vapour. This has been discovered in two ways. The first is by spectroscopy in the near-infrared and beyond. The spectroscopes are either ground-based or on Mars probes such as Mariner 9, Viking, Mars Global Surveyor and Mars Express, and Venus probes such as Venera and Pioneer Venus. The second method involves direct *in situ* mass spectroscopy analysis, by which a sample of the planetary atmosphere is collected and the nature of its constituents analysed, according to their atomic masses. That of the H_2O molecule, for example, with its two atoms of hydrogen (mass of each = 1) and one of oxygen (mass = 16), is 18. If the instrument detects a strong signal at atomic mass 18, the presence of water is indicated. Quantitative analysis is a delicate process, however, as there are other substances which may affect the 18 reading – one being the $^{15}NH_3$ isotope of ammonia. This method has been used successfully – a notable example being the detection of water vapour within Halley's comet by mass spectrometers mounted on the Giotto and Vega spacecraft.

In later chapters we shall see how the ISO detected water vapour in the stratospheres of the giant planets and in several comets. Furthermore, during the 1980s water ice was detected in the rings of Saturn and on the Galilean

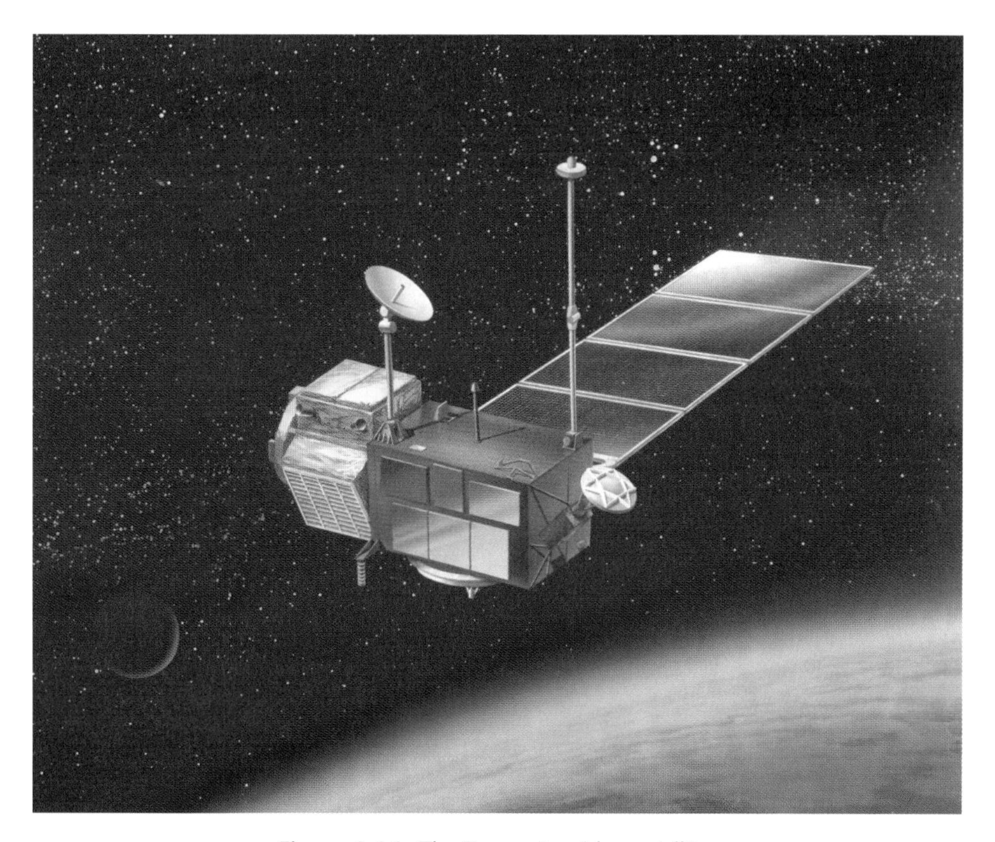

Figure 1.16. The Topex–Poseidon satellite.

satellites of Jupiter, via its spectral signatures in the near-infrared between 2 μm and 3 μm.

The search for ice on Earth

Obviously, we do not need to look far to find water on Earth. Indeed, our knowledge of its abundance, in solid and gaseous forms, and its local and seasonal evolution, are important to our understanding of the behaviour and evolution of Earth's climate.

The remote sensing techniques employed by astronomers also have their uses for climatologists: water-vapour spectroscopy, especially in the radio and millimetric domains, is carried out from ground stations, and from satellite platforms looking down at the Earth. These observations are augmented by others taken *in situ* at observatories on the ground, where constant hygrometric measurements are made of the humidity of the air (the percentage of saturation, which is the ratio of actual vapour pressure to maximum pressure corresponding to the temperature of the atmosphere), and of the amount of water vapour in the atmosphere. Such measurements are vital not only to geophysicists but also to astronomers: the less water vapour at a site, the better it will be for astronomical observations.

With their interest in climatic evolution, geophysicists also study the world's oceans. From 1992 until 2006 the Topex-Poseidon satellite – a joint venture by NASA and the CNES – constantly measured sea levels to an accuracy of the order of a centimetre. Its successor – PICASSO-CENA – will be launched in the near future.

Water in its solid form has been the object of study by glaciologists. Core sampling in the Arctic and Antarctic, to depths of up to 3 km, has provided unique data concerning isotopic ratios in the palaeo-atmosphere of the Earth, opening a window on the past evolution of the planet over millions of years.

2
The quest for cosmic water

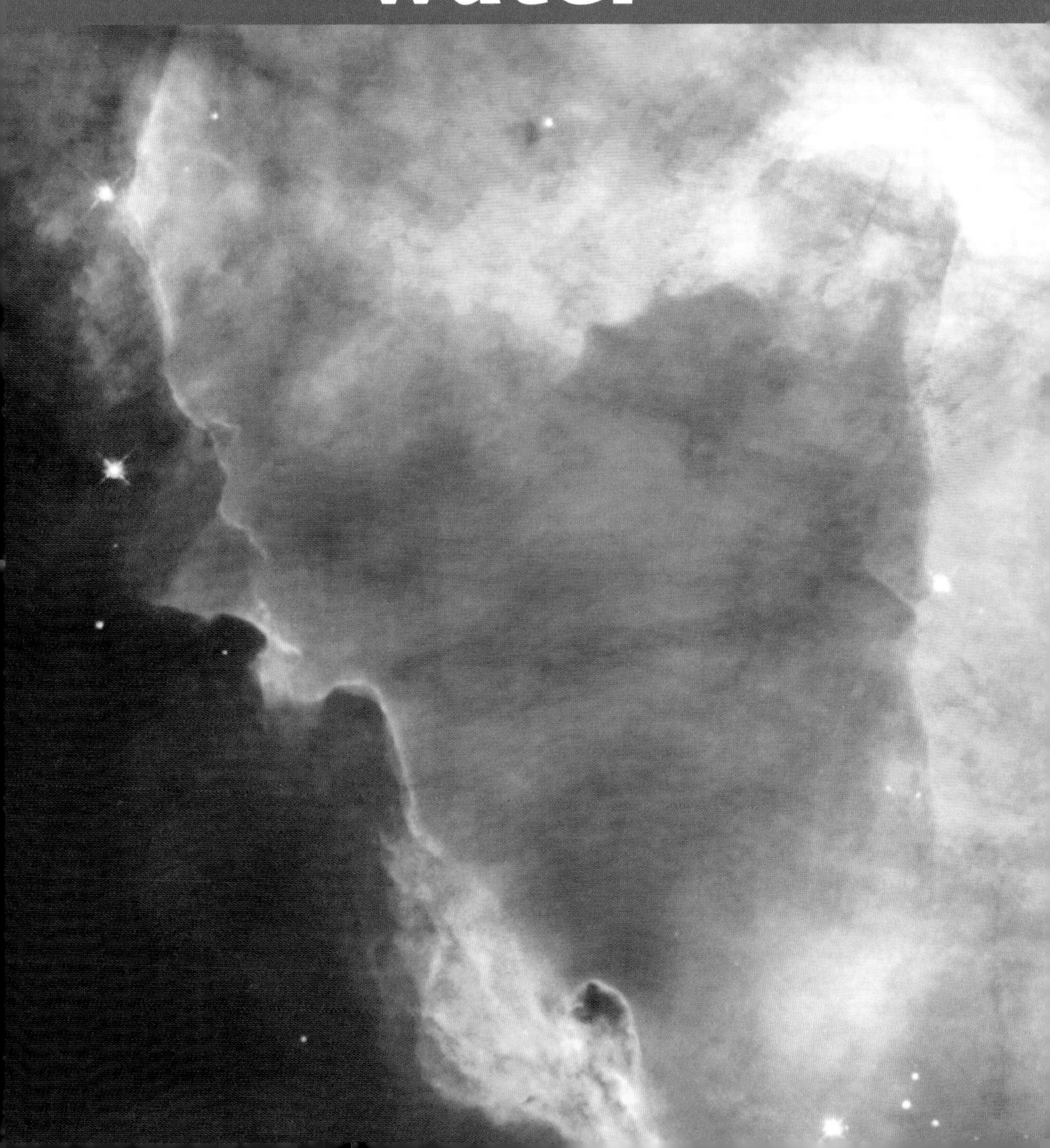

In 1877 the Italian astronomer Giovanni Schiaparelli announced the presence of *canali* on Mars. The word means ‘channels’, but was translated as ‘canals’ – leading to the belief that they were artificial in origin and had been built by intelligent beings. From that time onwards, the quest for water in the cosmos has continued unabated. Between 1995 and 1998, the memorable ISO mission revealed that water is to be found everywhere around the Universe, from nearby planets to the most distant galaxies, whether in the form of water vapour or ice. Other space missions follow the trail of ISO, among them Spitzer, SWAS and Odin, and soon Herschel, to be launched in 2008.

People of all civilisations have wondered about the probability of extraterrestrial life. Philosophers and scientists of the seventeenth and eighteenth centuries gave this question, dating from antiquity, a new impetus. For example, Cyrano de Bergerac, in *Histoire Comique des Etats et Empires du Soleil et de la Lune,* told of an encounter with the inhabitants of the Moon – the Selenites. Again, Fontenelle, in *Entretiens sur la Pluralité des Mondes,* imagined the existence of inhabitants of nearby worlds. And in the nineteenth century, astronomer Camille Flammarion reopened the debate in his book *La Pluralité des Mondes Habités*.

How are we to detect signs of extraterrestrial life? Using the analogy of the Earth’s history, astronomers have begun the search by looking for liquid water – firstly on nearby planets like Mars, and then on other more distant bodies in the Universe.

In this chapter we shall outline the principal stages of the quest for extraterrestrial water.

1877: CANALS ON MARS

In the late nineteenth century, observations of Mars revealed ‘canals’; and some observers thought that they were not natural features (Figure 2.1). These were the famous *canali* discovered by the Italian astronomer Schiaparelli in 1877. Certain well-known astronomers – such as the American Percival Lowell – had no hesitation in claiming that these ‘canals’ were indicators of life on Mars. Of artificial origin, they thought, they had been built by intelligent beings battling against drought. But they are merely an optical illusion – a product of lower-resolution visual observations (Figure 2.2). A heated debate – ‘Mars fever’ – ensued between supporters of the ‘canal’ theory, headed by Lowell, and its detractors, grouped behind Eugène M. Antoniadi. The controversy split the two camps for decades, but also stimulated much research into the possibility of water on Mars.

Then, in the first half of the twentieth century, the burgeoning science of spectroscopy led to a search for the signatures of water in the visible and near-infrared domains.

On Mars, oxygen, with its transition at around 0.7 μm, was the first target.

The Eta Carinae Nebula (NGC 3372).

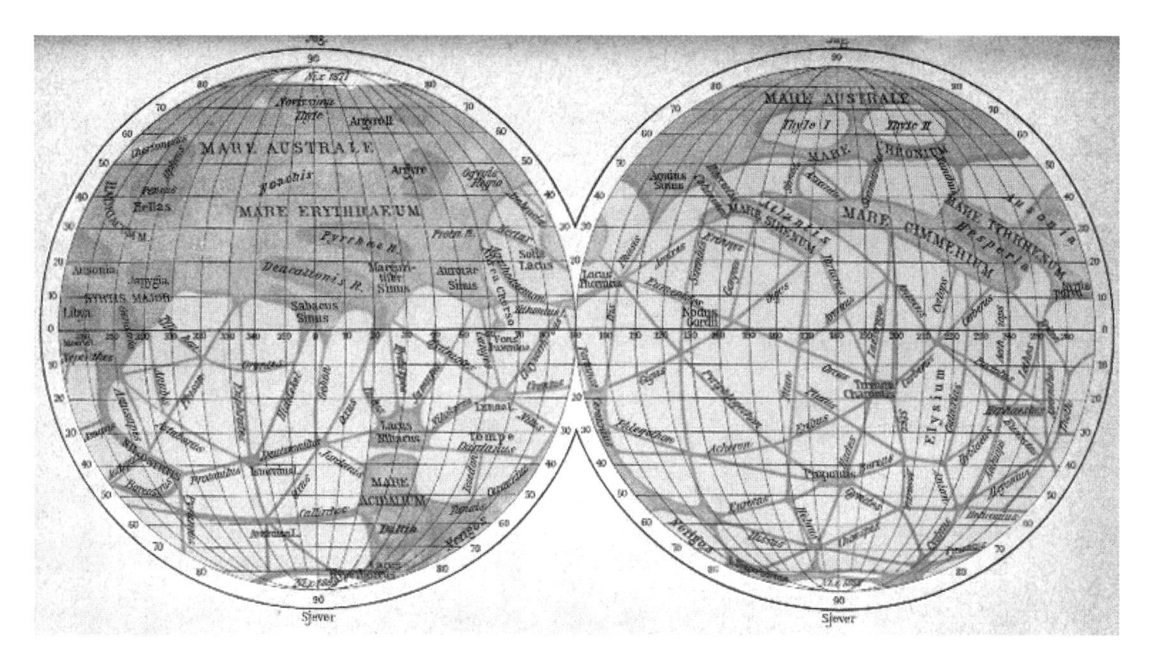

Figure 2.1. Map of Mars by Schiaparelli, 1879. The rectilinear structures are the so-called 'canals', considered by others to be signs of intelligent life.

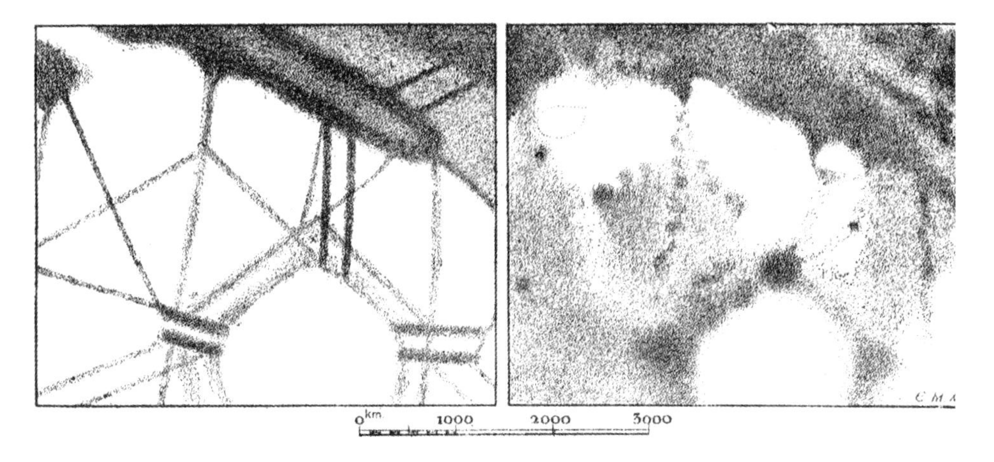

Figure 2.2. The Elysium Planitia region of Mars: a comparison between Schiaparelli's observations (1877–90) and those of Antoniadi (1909–26). It can be seen that, with improved spatial resolution, the 'linear' structures become irregular, contradicting the interpretation of martian canals. (From E.M. Antoniadi, *La Planète Mars*, Burillier, 1930.)

Then came the search for CO_2, at 1.6 and 2 μm. Results were at first negative, until the detection, in 1947, of traces of CO_2 by the American astronomer Gerard Kuiper. No trace of water was, however, detected in the martian atmosphere, in spite of the efforts of French astronomer Audouin Dollfus, using stratospheric balloons.

1950–1970: MARS, SATURN AND INTERSTELLAR WATER

In 1965 the first images from the Mariner 4 probe proved that the martian 'canals' do not exist. This put paid to the myth of life on Mars, and the hypothesis was laid to rest. The martian polar caps had long been attributed to the presence of water ice. Data from spaceprobes from the 1960s onwards showed that although water ice was a partial constituent, the caps were mostly covered with frozen carbon dioxide.

Ice in the solar system

Extraterrestrial water ice is more easily detected than water vapour, as the presence of the Earth's atmosphere is unfavourable to the detection of the latter. Water ice, however, presents spectroscopic bands which, in clear skies, can be observed from the Earth's surface. This is particularly true of the two bands at 1.5 μm and 2 μm, well away from the signatures of gases in the terrestrial atmosphere. During the second half of the twentieth century, astronomers turned this to their advantage, and from the 1960s onwards water ice was identified, using near-infrared spectroscopy, in various solar system objects – notably in Saturn's rings, and in three of the Galilean satellites of Jupiter, Europa, Ganymede and Callisto. Then followed further discoveries of water ice, mostly upon the surfaces of Saturn's moons (with the exception of Titan, which is cloaked in a cloudy, dense atmosphere).

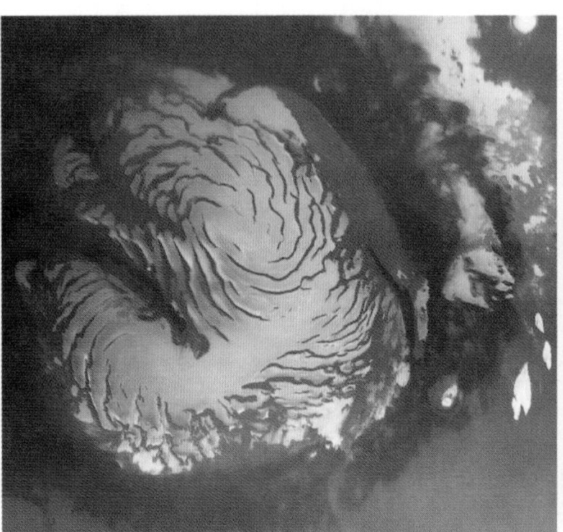
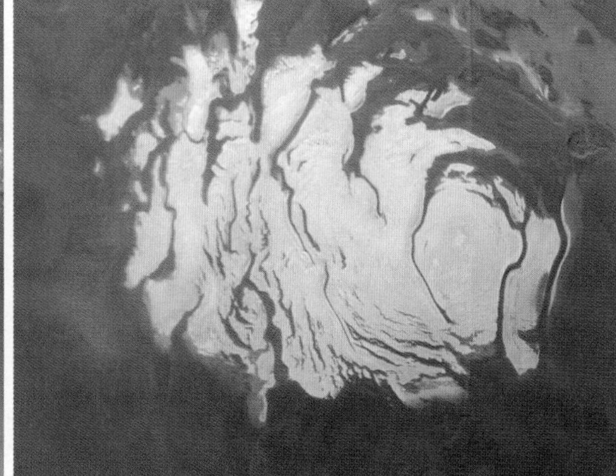

Figure 2.3. The martian polar caps. At both the north (a) and south (b) poles, several thousand metres of ice, sediment and dust have accumulated. Around the north pole these deposits are composed almost entirely of a layer of relatively pure water ice. At the south pole the deposits are partly covered by a permanent cap of solid, very reflective carbon dioxide. Observations by Mars Express showed that the carbon dioxide layer covers an even greater quantity of water ice.

Interstellar water and radio astronomy

Water exists as ice in the solar system, but is this the case elsewhere in space? Although Earth's atmosphere presents an obstacle to the detection of water vapour, we might expect to be able to detect ice in cosmic objects by means of ground-based observations. However, the spectral signatures of water-ice in the near-infrared (such as those observed in Saturn's rings and the satellites of Jupiter) are too weak to be observable on distant objects. More marked signatures (at 3 μm and in the far-infrared around 45 μm) are masked by absorption in the atmosphere. Thus it was that the first measurements of water in the Universe, from the ground, were of its gaseous phase.

In 1955, the American astronomer Charles Townes had already predicted the presence of the water molecule in the interstellar medium, as well as those of atomic hydrogen (H) and the OH radical, and the NH_3 and CO molecules. These predictions were a consequence of the efforts of other pioneers in the field of radio astronomy – among them Jansky, Reber, Oort, Van de Hulst and Ewen. Work begun in 1951 had led to the discovery of the presence of neutral hydrogen, with its very weak 'hyperfine' transition signal at 21 cm (1,420 MHz). Even though this transition is fairly improbable, its detection was made possible by the fact that atomic hydrogen is the most abundant element in the interstellar medium, observable at very great distances.

These radio observations were made using *heterodyne* spectroscopy – a method whereby measurement is possible on a reduced spectral band (of about 0.5 GHz) with much enhanced resolution (of the order of 10^6 times), allowing absolutely unambiguous identification of the spectral line in question. Moreover, the considerable size of radio astronomy antennae, often working together as interferometers, guarantees very good spatial resolution, thereby limiting the size of objects to be observed.

Building on Townes' work, Barrett and Weinreb detected the OH radical in 1962, at a frequency of 1,650 MHz (λ = 18 cm). The 18-cm OH emission produces several distinct lines with an intensity and width that can be measured. These parameters provide us with an insight into the processes of excitation of the radical, and the physical conditions pertaining in its environment. In the case of the OH radical, the lines appear very intense and narrow. To explain this effect, astronomers have proposed a signal amplification mechanism: the *maser* (Microwave Amplification by Stimulated Emission of Radiation) effect (by analogy with the laser), which amplifies radiation and is mainly associated with visible light. The maser process, attributed to the cosmic radiation field, serves to increase the number of levels of excitation of the OH radical and creates strong emission lines. Although this effect favours the detection of interstellar components, it makes it difficult to measure their abundances.

Hydrogen (H) and the OH radical had been detected in the interstellar medium. Now came the turn of water vapour itself. In 1969, Townes' team identified water vapour in star-birth regions. The presence of water was inferred from a specific emission at a frequency of 22 GHz (λ = 1.35 cm). The intensity of

Radio interferometry

In its simplest form, a radio interferometer consists of two antennae aligned along an east–west axis, simultaneously observing the same celestial body. By combining the signals received by each antenna, interference fringes are created due to the difference in the arrival time of two beams arriving at each antenna. As the observed source follows its diurnal path in the sky, this difference varies, and the evolution of the system of fringes provides information about the dimension of the source along its east–west axis. Interference fringes can also be observed using a delay line, artificially introducing a difference in the arrival of the beams. The greater the distance between the two antennae, the more accurate will be the measurement of the diameter of the source.

The technique can be improved by using a whole array of antennae, along both north–south and east–west axes simultaneously. Radio sources observed in this way can be mapped. The Very Large Array, in New Mexico, USA, is particularly powerful, possessing twenty-seven antennae, each 25 metres across, and is capable of spatial resolution better than 1 arcsecond at frequencies up to 50 GHz ($\lambda = 6$ mm). The IRAM interferometer at Plateau de Bure, near Gap, France, has six antennae, each 15 metres across, with spatial resolution of the order of 1 arcsecond at frequencies up to 260 GHz ($\lambda = 1{\cdot}1$ mm).

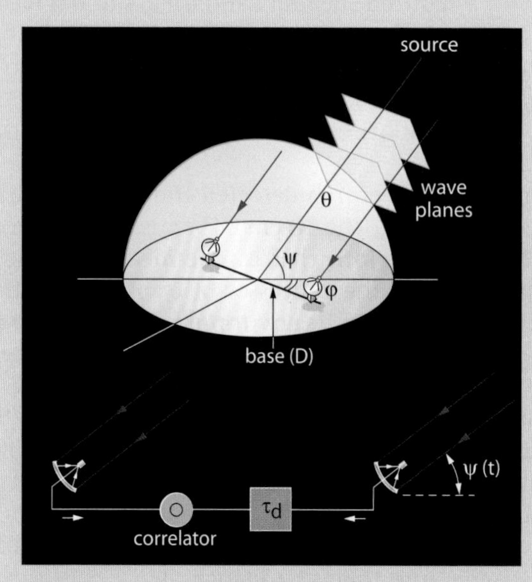

Figure 2.4. The principle of the radio interferometer. The two antennae, on baseline D, point in direction θ towards the source. Angles ψ and φ defining θ depend on time, because of the diurnal movement of the source. This movement creates a phase difference between the two incoming beams. The phase difference may also be electrically generated by a delay line τ_d. The signals from the two antennae are combined within the correlator. (From P. Léna, *Méthodes Physiques de l'Observation*, InterEditions-Editions du CNRS, 1986.)

this transition is relatively weak, but Earth's atmosphere is not totally opaque at this wavelength, and it may therefore be observed: the signal of water in the vicinity of young stars still forming. The detected water molecules are in the discs of gas and dust surrounding hotter central regions, where stars are in the first stages of their lives. However, the quantity of water present in these regions cannot be determined by these observations. As in the case of the OH radical, it is the maser effect amplifying the signal which facilitates detection of the H_2O transition, as certain energy levels of the molecule are excited due to the presence of cosmic radiation and nearby stars. For this reason, H_2O emissions are often detected in association with OH sources.

Water is therefore definitely a feature of the cosmos, although it is still not easy to estimate its abundance.

1970–1990: MARS AND THE COMETS

During the 1970s, observations revealed that water is present on Mars only in tiny quantities - so much so that it had been undetectable with the instrumentation available in the first half of the twentieth century. Spectroscopic measurements by the spaceprobes Mariner 9, Viking 1 and Viking 2 - orbiting Mars, respectively, in 1971 and 1975–80, were finally able to provide exact data concerning the quantities and seasonal evolution of water vapour on Mars.

Other techniques were later employed, including spectroscopic measurements by Mars Global Surveyor (launched in 1996) and Mars Express (launched in 2003). Mars Odyssey (launched in 2001) used a neutron spectrometer.

Comets are smaller, wandering bodies usually less than 10 km in diameter, and they pursue very eccentric orbits. Due to their low-temperature environment, and the fact that they suffer few collisions, they have remained practically unchanged since their origin. For this reason they have always been intriguing objects to scientists, who regard them as valuable evidence of what conditions were like in the early solar system. From 1950 onwards, the American astronomer Fred Whipple predicted, using thermochemical models, that comets would be composed mainly of water ice. However, this was not easily verified.

Observers studying comets from the Earth encounter many problems. When a comet is close to the Earth (and therefore close to the Sun), it may be bright enough to be observed; but gas and dust ejected from the coma, as a result of the sublimation of ices at the surface of the cometary nucleus, obscure the nucleus itself. In principle, the water vapour might be observable from the Earth; but again, the presence of our atmosphere creates difficulties (see fact box, p. 24). Conversely, when a comet is far away its temperature is so low that sublimation is suppressed, and the nucleus is exposed; but so little light is reflected from it that it cannot be seen.

Unable to directly observe the water, astronomers have concentrated on the OH radical, which is a product of the dissociation of water by the solar flux. In 1973, at the Nançay radio astronomy facility in the Cher *département* of France, astronomers Eric Gérard, François Biraud and their colleagues detected the presence of the OH radical in comet Kohoutek. Their observations were of great value in determining the physical parameters of cometary environments; for example, temperature, and outgassing velocities. There followed a systematic programme of observations of the OH radical in comets, providing a database on a great number of objects. Even before the first detection of H_2O in comets, the identification of the OH radical – comparable in abundance with that of atomic hydrogen (H) – had provided valuable clues as to the presence of large quantities of water in comets. For the last thirty years the rate of OH production in comets has been used as an indicator of their outgassing of water (see Chapter 4).

Water was directly detected in a comet in 1986, when new observational methods were combined with a fleet of five spaceprobes to study comet Halley. Whipple's hypothesis was brilliantly confirmed. To be more precise; in autumn 1985 a team led by Michael Mumma first detected, in a seminal experiment, the emission lines of water vapour. The lines were observed in the fundamental spectroscopic bands at 2.7 μm, from the Kuiper Airborne Observatory (KAO), with its 90-cm telescope, flying at an altitude of 14 km.

1994–1995: WATER VAPOUR AND THE GALAXIES

In 1994, millimetric radio astronomy scored a remarkable success: the detection of water vapour in a very distant galaxy, IRAS 10214+4724, 1.5 billion light-years away. This was achieved by Fabienne Casoli and colleagues at the Paris Observatory, using a method based on redshift: the further away a galaxy is from us, the faster it appears to be receding (see fact box, p. 40). Velocity of recession is proportional to distance, and radiation arriving from these sources is shifted towards the red (lower-frequency) end of the spectrum because of the Doppler effect. This redshift is directly proportional to the speed of recession (Figure 2.6). This property of distant sources has a very important consequence for the observation of the spectral lines of water vapour: a transition occurring in the submillimetric domain (totally unobservable from Earth) can, because of the Doppler effect, be measured at a lower frequency in a domain observable from the ground (Figure 2.7). It was in this way that the French team, using the 30-metre dish at IRAM (Pico Veleta, Spain), were able to observe an H_2O transition of original frequency 752 GHz (unobservable from Earth), Doppler-shifted to an apparent frequency of 229 GHz, within an atmospheric window (Figure 2.5). This was the first observation of water vapour in a distant galaxy. This result – alongside the discovery of OH, CO and other molecules in external galaxies – shows us that our Galaxy is not the only one in which such molecules have been synthesised. Their formation seems to have occurred quite early in the cycle of galactic evolution.

Observation of the maser transition at 22 GHz has led to studies of water vapour in about ten extragalactic objects. In 1995 a central torus of gas and dust was detected in the galaxy NGC 4258 – an observation made possible with the

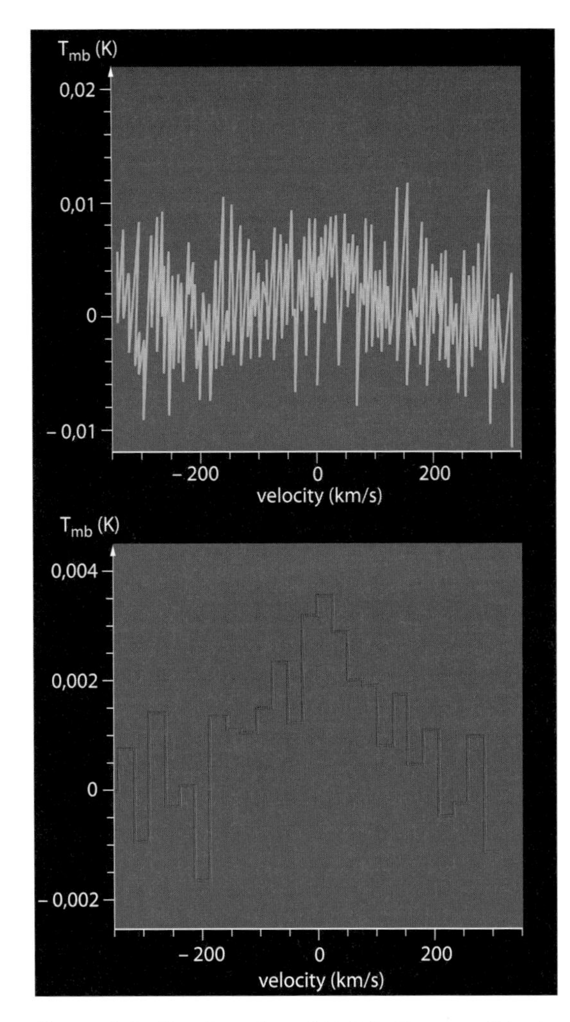

Figure 2.5. The detection of H_2O in the very distant galaxy IRAS 10214+4724. (Above) the raw spectrum at high resolution; (below) the spectrum at intervals of 10 km/s, showing the emission. The galaxy's Doppler shift corresponds to $z = 2.3$, signifying that the light we see today was emitted 1.5 billion years ago. The observed line corresponds to an emission at a frequency of 752 Ghz ($\lambda = 399$ μm), which cannot be observed from the ground. However, the Doppler shift of the receding source, due to the expansion of the Universe, shifts this transition towards the red, to a much lower frequency ($\nu = 228.86$ GHz, which is observable from the ground). This spectrum represents the first detection of water vapour in a very distant galaxy. (From F. Casoli *et al., Astron. Astrophys.*, **287**, 716, 1994.)

very high resolution (<10 parsecs – <32 light-years) achieved by interferometric methods, and a much enhanced (10^6) spectral resolving power. This torus – with a radius of only 0.1 parsec – rotates around a supermassive black hole with an internal mass of 40 million Suns. The density of this object is one of the greatest ever measured for a black hole at the centre of a galaxy.

1995–1998: THE INFRARED SPACE OBSERVATORY

There is only one way to guarantee an unobscured view of extraterrestrial water: by observing from above the Earth's atmosphere, which is opaque to all the spectral signatures of water vapour. It is therefore necessary to send into Earth-orbit a telescope equipped with spectroscopic instruments 'seeing' the whole range of infrared radiation.

Thus was born the Infrared Space Observatory (ISO), commissioned by the European Space Agency (ESA) in the early 1980s, and finally launched atop an Ariane 4 rocket, from Kourou, French Guyana, on 17 November 1995. The ISO carried a 60-cm telescope, cooled to a temperature of 8 K, and was equipped with four other instruments with detectors working at a temperature of almost 2 K – extremely low temperatures that ensure maximum sensitivity in

Redshift in radiation from distant sources

The observations of the American astronomer Edwin Hubble showed that galaxies are, in general, moving away from us. The farther away they are, the faster they appear to be moving. Hubble demonstrated that their velocity of recession is proportional to their distance, and the ratio between the two is the Hubble constant – about 70 km/s/Mpc.

This recession of the galaxies was revealed by the redshift of spectral lines of the objects observed. What is this phenomenon? It is analogous to what we hear when we stand by a railway, and a train approaches us at speed, and then moves away from us. The sound of the approaching train is high-pitched, then lower-pitched as it recedes. This is the Doppler effect. This shift in frequency occurs at all frequencies (or wavelengths), whatever the distance of the object in question, as long as it is motion relative to the observer.

If galaxies are very far away, the shift caused by the Doppler effect is very considerable. It is measured by the parameter $z = \Delta\lambda / \lambda_0$. If the radial velocity is small, then $z = v/c$, where v is velocity of the celestial body and c is the velocity of light in a vacuum (300,000 km/s). In the contrary case, the complete equation is written: $1 + z = (c + v) / (c - v)^{1/2}$.

This equation shows that v tends towards c as the spectral shift tends towards infinity. In the millimetre domain, astronomers have been able to measure, for example, an H_2O transition of initial frequency 752 GHz, redshifted (shifted towards lower frequencies) to 229 GHz, observable from the ground. Nowadays it is possible to measure spectral shifts of as much as $z = 7$.

Nearer to home, planets and comets within the solar system also show a (slight) Doppler shift as a result of their revolution around the Sun, its amplitude being much lower (of the order of a few tens of km/s^2). Nevertheless, observers have used it to their advantage many times in studies of the spectral lines of water, in order to separate those of the object observed from those of terrestrial origin. It was in this way that the first identification of H_2O in Halley's comet was carried out, using the Kuiper Airborne Observatory.

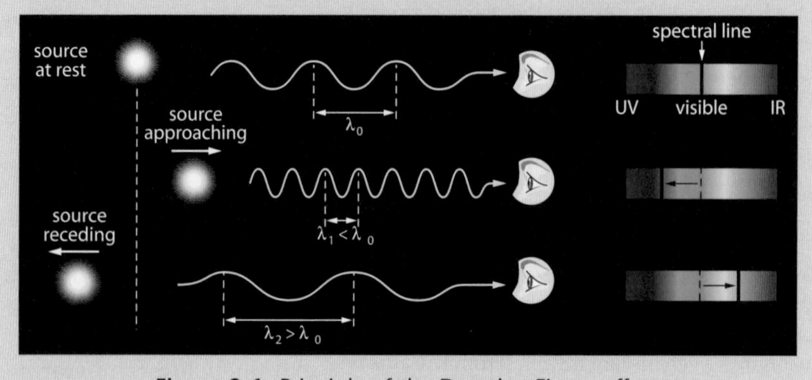

Figure 2.6. Principle of the Doppler–Fizeau effect.

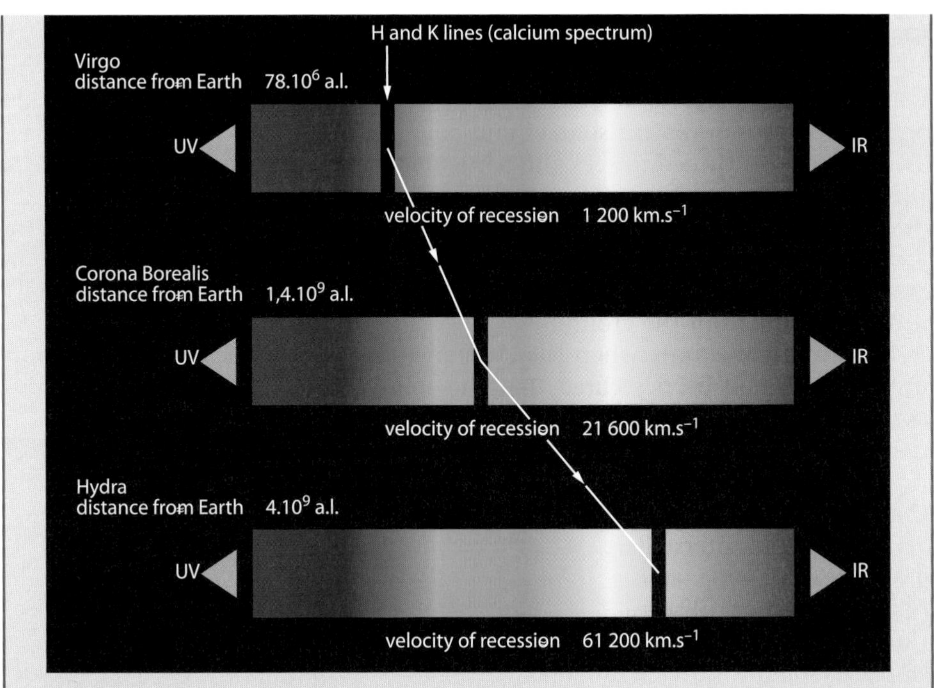

Figure 2.7. The Doppler–Fizeau effect. Spectral lines normally unobservable from the Earth may be measured at lower frequencies, in a spectral domain observable from the ground, as a result of the recession of the galaxies.

the measurements. The whole assembly sat within a huge cryostat cooled with liquid helium – the lifetime of the mission being determined by the rate of evaporation of the helium into space. Originally supposed to function for eighteen months, the ISO remained active for 2½ years, and its scientific operations ceased in May 1998.

The four instruments were a camera, a photometer and two spectrometers – the latter of which were used in the detection of water by the identification of the spectroscopic signatures of water vapour and ice. They possess nothing like the same resolving power as is found in heterodyne spectroscopy (an optimum ratio might be $3{:}10^4$), but the domain covered was wide (2–200 μm). This spectral range allowed the simultaneous observation of many different transitions, guaranteeing unambiguous identification of the molecules being sought.

The results from the ISO marked a real revolution in our knowledge about water in space. The satellite demonstrated that water is omnipresent in the cosmos, from nearby planets to the most distant galaxies, whether in the form of water vapour or ice.

Water among the stars

Ground-based observers detect the spectroscopic signatures of ice only weakly, and, in the case of distant sources and comets, almost not at all. Nevertheless, astronomers continued to seek the particularly intense signature of water ice in the infrared at 3 μm – preferentially within the cool, dense clouds of the interstellar medium – and several positive results were recorded. In the 1990s the Kuiper Airborne Observatory (KAO), flying at an altitude of 14 km and therefore above most atmospheric absorption, detected interstellar water ice at 45 μm – a feat impossible from the ground.

However, it was only when the ISO's measurements, spanning the whole of the infrared, had been secured that the spectroscopic signatures of water ice could be studied and its nature (amorphous or crystalline) determined. Ice that has formed on the surfaces of cold interstellar grains tends to be amorphous, but ISO observations had a surprise in store: crystalline water ice was found in the vicinity of certain young (and some late-stage) stars (Figure 2.9). These results suggest a temperature limit beyond which crystallisation occurs.

As for water vapour: it has been detected on innumerable occasions. The ISO confirmed its presence in old, Mira-type stars, where astronomers had already detected H_2O masers. ISO spectra showed emission lines of water vapour, involving an excitation temperature of 500–1,000 K, and also a component associated with dust. These measurements suggest the existence of a warm molecular envelope around late-stage stars.

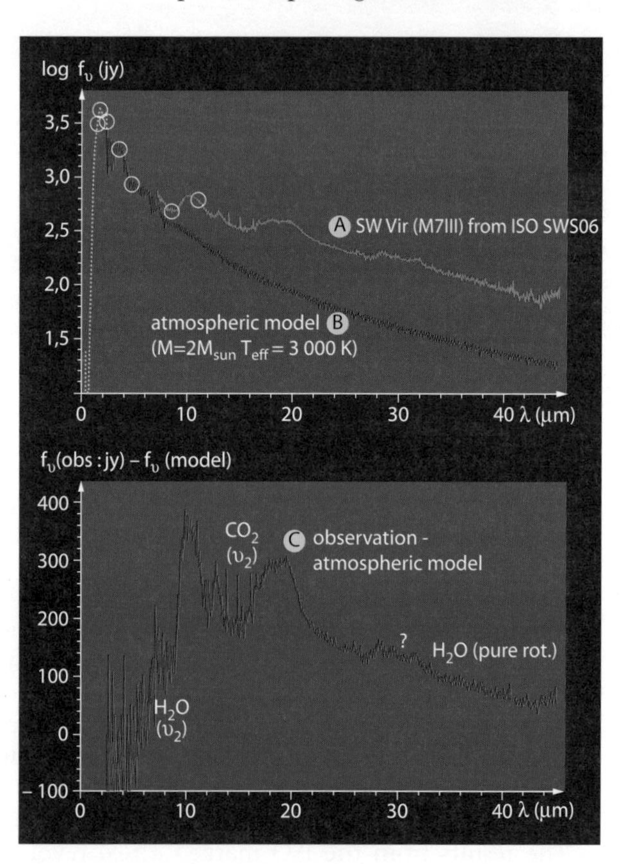

Figure 2.8. The detection of water in late-stage SW Virginis and M-type giant stars, by the ISO-SWS spectroscope. (A) observed spectrum; (B) theoretical spectrum calculated by an atmospheric model; (C) difference between A and B revealing H_2O emission lines. (From T. Tsuji *et al.*,1999.)

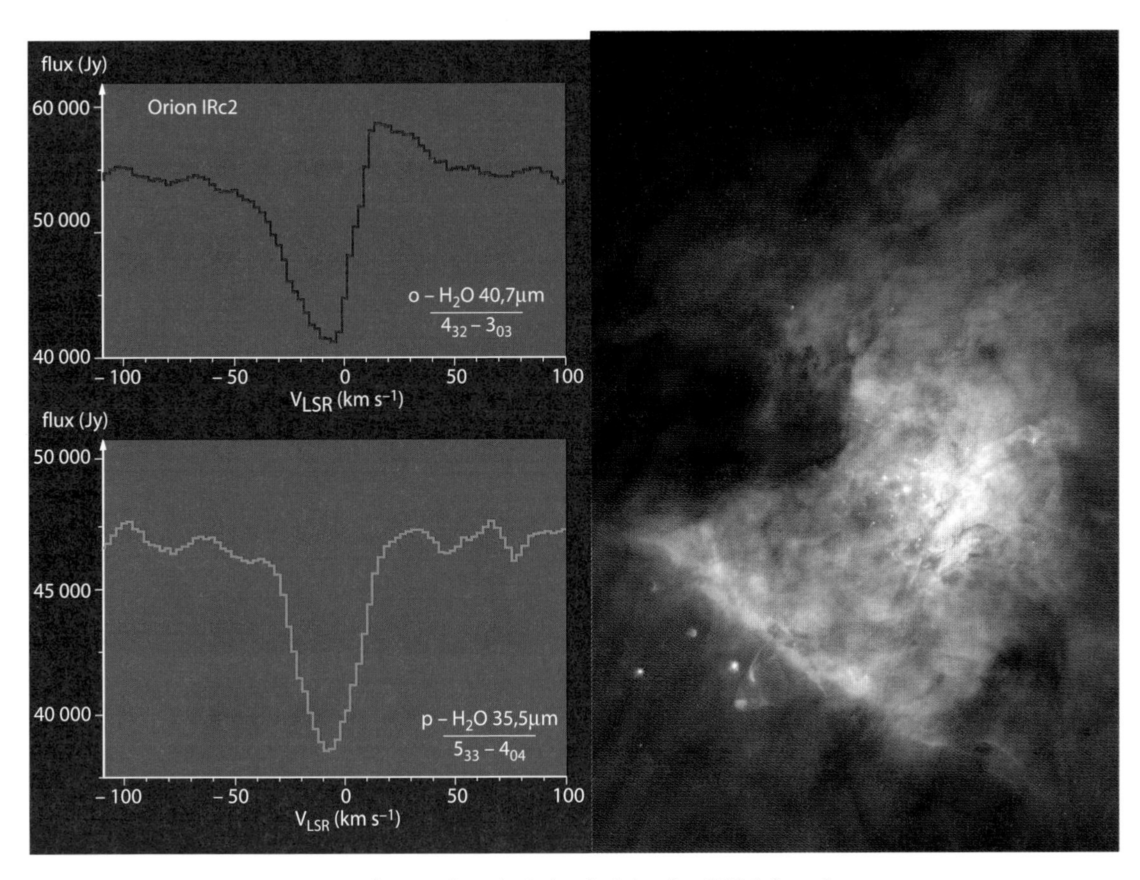

Figure 2.9. Detection of water lines in Orion Irc2 by the SWS infrared spectroscope on the ISO satellite. (From Van Dishoek *et al.*,1999.)

The ISO also detected water vapour at the other end of the stellar-lifetime scale, in star-forming regions. This was another confirmation of earlier results involving H_2O masers in the radio domain. The extension of spectral analysis into the whole of the infrared domain has facilitated the determination of the physical parameters of the clouds which are the seat of these emissions – in particular, the excitation temperature, which may be as high as 1,000 K. The abundance of H_2O compared with that of hydrogen may vary by several orders of magnitude! In well-known intense sources such as the Orion Nebula or the Galactic Centre (Sgr B2), the relative abundance of H_2O is of the order of 10^5 or 10^4.

It is worth mentioning, in passing, that measurements of heavy water (HDO) also provide vital information about the temperature at which the H_2O molecule forms in the interstellar medium. In fact, comparative (ISO) measurements of H_2O and its HDO isotope (detected from the ground using heterodyne submillimetric spectroscopy) have allowed us to determine the D:H ratio within the warm central areas of star formation. They show a deuterium excess of 4–20

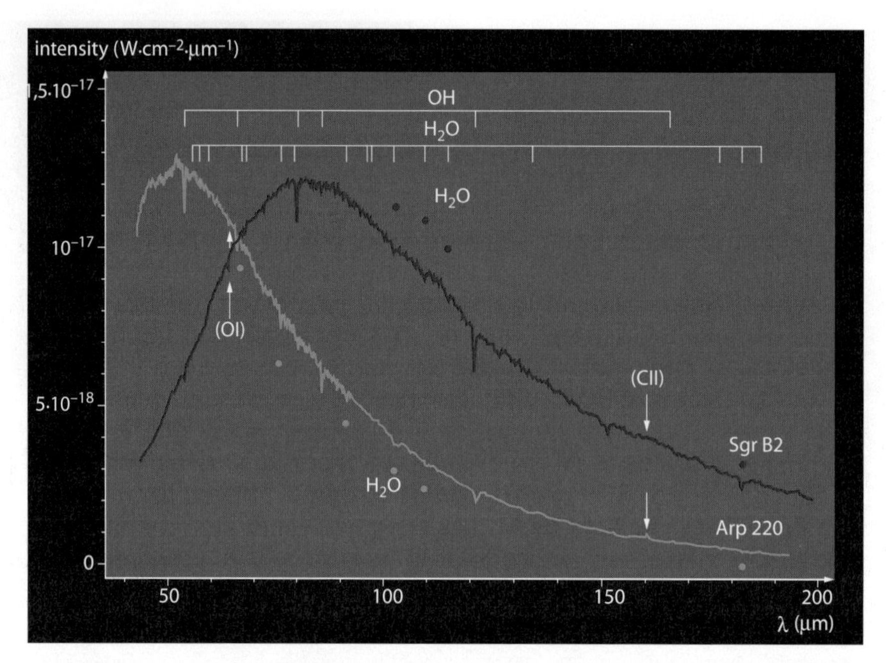

Figure 2.10. Water vapour in distant galaxy Arp 220. This spectrum was obtained in the far-infrared by the LWS spectrometer onboard the ISO satellite. The spectrum of the Galactic Centre object Sgr B2 is shown for comparison. It can be seen that the two spectra exhibit absorption lines due to OH and H_2O. (From P. Cox *et al.*, ESA SP-427, 631, 1999.)

times compared with values measured in the primitive solar nebula (2:10^5). Such an excess might be expected in the interstellar medium, where low-temperature ion–molecule and molecule–molecule reactions promote deuterium excess of H_2D^+ (and thence HDO), and of HDO itself. Examples of these reactions are (ion–molecule) $H_3^+ + HD \rightarrow H_2D^+ + H_2$ and (molecule–molecule) $H_2O + HD \rightarrow HDO + H_2$. How can it be, then, that the excess also occurs in warm, star-forming regions? It could be that it has already happened on the surface of cold grains when the warm stellar cores shrank inwards, and the enriched material was preserved during the sublimation phase. We shall see, in the ices of the solar system, similar evidence of increased deuterium – the signature of their formation at low temperatures.

In the case of external galaxies, the ISO confirmed and broadened H_2O-maser results obtained by radio astronomy. Several rotational lines of water vapour have been detected, with variable levels of excitation, in some nearby or ultra-luminous galaxies such as Arp 220 (Figure 2.10).

Water in the solar system

Lastly, we should mention the ISO results concerning water in the solar system. One unexpected development was the discovery of water vapour in the

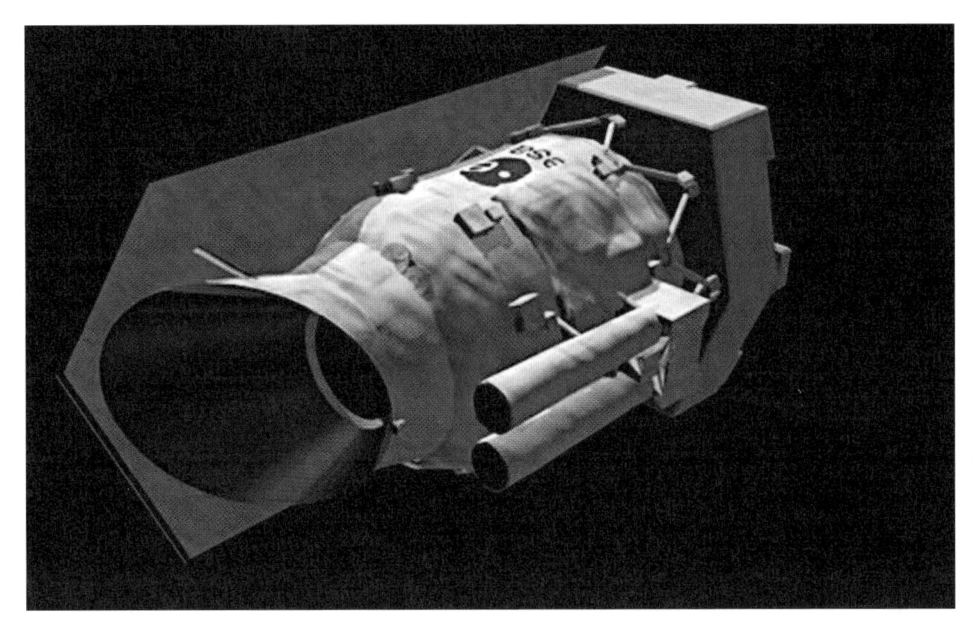

Figure 2.11. The Infrared Space Observatory. Launched on 17 November 1995, it represented a revolutionary step forward in our search for extraterrestrial water. Water is omnipresent in the cosmos – from nearby planets to the most distant galaxies – as water vapour or ice.

stratospheres of the giant planets. Given the low temperatures at a lower altitude, the source of this water vapour must be external. It is the signature of an incoming oxygen flux, perhaps linked to the micrometeoroids present throughout the outer solar system (discussed in Chapter 5).

Although water vapour had been detected in a comet in 1985, the ISO made the first definite measurements of the presence of water ice. Of course, the comet has to be far away from Earth for its water to be in the form of ice, so a sensitive instrument is needed. In 1997 the ISO detected water in a distant but very active comet - Hale–Bopp - and also secured spectra of unrivalled accuracy showing water-vapour emission lines (further discussed in Chapter 4).

THE POST-ISO ERA

The SWAS and Odin missions

Two other space missions dedicated to the quest for water in the Universe took up the baton from the ISO. Both were Earth orbiters. The first of these missions - the American Submillimeter Wavelength Astronomical Satellite (SWAS) - carried a 60-cm telescope, and was launched in 1999. The second probe was the European Odin, with a 1-metre telescope, launched by Sweden, in collaboration

Figure 2.12. The Submillimeter Wavelength Astronomical Satellite.

Figure 2.13. The Odin satellite.

with Canada, Finland and France, in 2001. These satellites were more modest in their aims, and had more limited vision than ISO. They worked with just a few spectral channels, centred around specific frequencies. These correspond to intense H_2O transitions of 557 GHz (a wavelength of 539 μm), and O_2 transitions (for Odin, 118 and 487 GHz, wavelengths 2.5 mm and 616 μm; for SWAS, 487 GHz only). The SWAS mission is astronomical, while Odin spends half of its time studying the Earth's atmosphere. The instruments carried are heterodyne spectrometers, working within a narrow spectral band around the frequency of the lines sought, with high resolution (Figure 2.14). The astronomical aspect is the

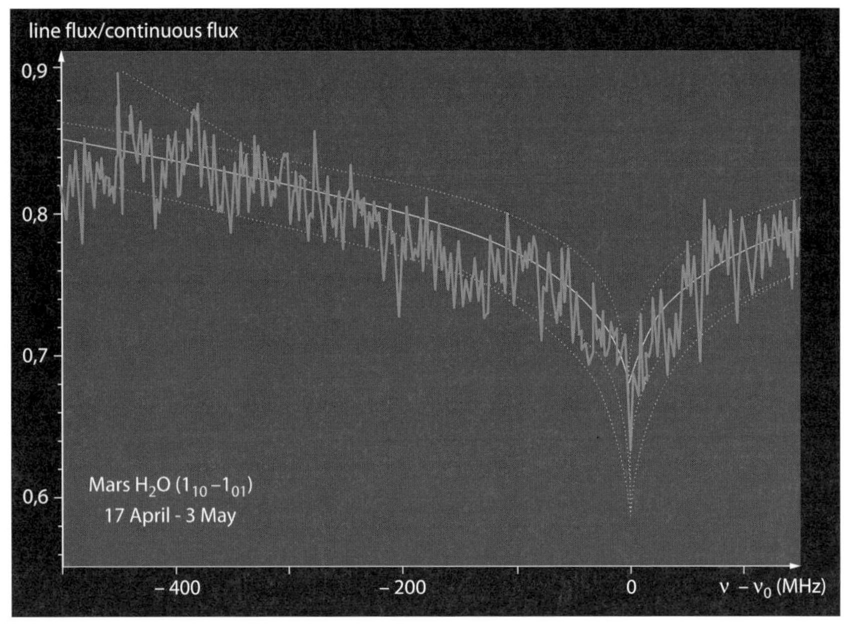

Figure 2.14. Detection of the water line at 556.935 GHz on Mars by the SWAS, compared with models of water distribution. This spectrum shows the vertical distribution of water vapour in the martian atmosphere. The dotted curves correspond to different theoretical models. The horizontal axis represents the difference between measured frequency and the central frequency of the line. (From M. Gurwell *et al.*, 2000.)

search for water, including the measurement of its abundance and the investigation of the role it plays in both solid and gaseous forms within the interstellar medium.

Another unresolved problem is that of the observed lack of abundance of molecular oxygen. To date, attempts to detect its submillimetric transition at 487 GHz have failed – suggesting that the abundance of molecular oxygen compared to the abundance of other well-known molecules (such as CO) is about fifty times less than chemical models predict. Greater knowledge about the abundance of H_2O in the various sources where water has been detected will help in determining amounts by mass of oxygen.

1999: water vapour in the Sun!

The water molecule is thermally stable up to temperatures of 3,000 K, so there was no expectation of detecting water vapour on the Sun's disc, as the temperature of the solar photosphere is around 5,600 K. However, sunspots, which appear darker, are considerably cooler, with temperatures of about 3,200 K. It is in these regions that water vapour has been found, using infrared spectroscopy at around 18 μm. The identification was carried out by comparison with a laboratory water-vapour spectrum involving a temperature of 1,820 K,

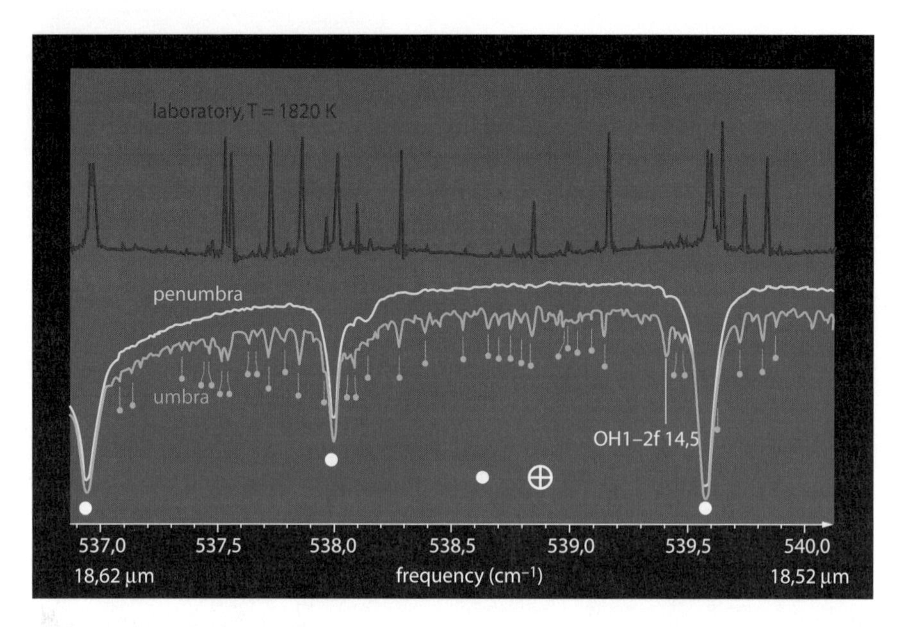

Figure 2.15. Water vapour in sunspots. Comparison of the infrared spectrum of a sunspot (umbral, in green) with a laboratory spectrum of H_2O at a temperature of 1,820 K (in red). The water vapour lines are very obvious. The spectrum taken within the penumbra (in yellow) at the edge of the sunspot, where the temperature is higher, shows no H_2O lines. More intense lines due to the Earth's atmosphere are present in both cases (white dots). (From T. Oka, *Science, 277*, 328, 1997.)

where a large number of rotational lines can be found which are present in the solar spectrum (Figure 2.15).

Where did the Sun's water come from? Its origin remains an open question. It may have been delivered by large numbers of captured Sun-grazing comets. Historically, several Sun-grazers have been observed; but SOHO – active since 1995 – discovers about a hundred of them every year. It has yet to be understood how the H_2O molecules survive within the sunspots without being destroyed by heat. This phenomenon is related to that of the water detected in late-stage stars with temperatures of about 3,000 K – again close to the thermal threshold beyond which the water molecule is doomed.

The Spitzer mission

Further space projects complement the work of the ISO. The American Spitzer Space Telescope (originally SIRTF) was launched in August 2003. It is a cooled 90-cm telescope, accompanied by highly sensitive infrared spectrometers and cameras. The Spitzer mission, with its great imaging and photometric capabilities, is particularly well adapted for cosmological studies and observing faint objects. However, its resolving power is less than that of the ISO, and it is less well suited for studying water vapour in space.

The Herschel mission

The Herschel Space Observatory (HSO) – formerly called FIRST (First Infrared and Submillimetre Telescope) – is the European Space Agency's fourth cornerstone project in its Horizon 2000 scientific programme, announced at the beginning of the 1980s.

Herschel is a space observatory working at far-infrared and submillimetre wavelengths. It is scheduled to be launched in 2008, and will carry a 3.5-metre telescope and an array of three other instruments: PACS, SPIRE and HIFI. PACS (Photodetector Array Camera and Spectrometer) and SPIRE (Spectral and Photometric Imaging Receiver) are imaging spectrometers covering, between them, the domain 60–250 μm. HIFI (Heterodyne Instrument for the Far-Infrared) is a heterodyne spectrometer, of very high spectral resolution ($R = 10^6$), working between 150 and 600 μm.

Figure 2.16. The Herschel spacecraft – an astronomical observatory working in the far-infrared and submillimetre domains. It is due to be launched in 2007.

The Herschel telescope will be cooled passively to about 80 K. The other instruments will be lodged within a superfluid-helium cryostat, using technology already utilised on the ISO mission. Herschel will be launched on an Ariane 5 rocket, and will orbit around Lagrangian point L2 – a point situated on the line between the Earth and the Sun at a distance of 1.5 million km from Earth. The lifetime of the mission will be at least three years.

Herschel is designed to study cold objects in the Universe – objects radiating principally in the infrared and submillimetre domains. In particular it will investigate any water vapour associated with these objects, so that there is potentially a vast range of targets: planets, comets, the interstellar medium, molecular clouds... all the way to super-luminous galaxies of distant cosmological origin. Herschel will be the first space observatory capable of observing the whole of the submillimetre spectral domain.

FUTURE PROJECTS: HERSCHEL AND SPICA

Two ambitious projects will follow SWAS, Odin and Spitzer: the Herschel satellite, designed and built by the European Space Agency, and the Japanese Space Agency's SPICA.

Herschel – to be launched in 2008 – will carry a 3.5-m telescope and three very sensitive spectrometers, supercooled for maximum sensitivity, to cover the whole of the far-infrared and submillimetre domains (see fact box, p. 20).

Further into the future, the even more ambitious Japanese SPICA mission will have the same objectives as Herschel, but its 3-metre telescope will be actively cooled, like that of the ISO. In this way, noise due to thermal emissions from the telescope will be restricted, ensuring much increased sensitivity.

3
The ice line and the birth of the planets

According to the 'primordial nebula' theory, the solar system formed from a cloud of gas and dust which, drawn inwards upon itself by its own mass, became a rotating disc. Water – present as ice and water vapour – was one of the prime constituents of this cloud. The 'ice line' – its position defined by the temperature of condensation of water and therefore at a fixed distance from the Sun – marked the boundary between the gaseous component of the protosolar material (the inner solar system) and the solid component (the outer solar system). It was one of the major factors in the differentiation of the planets into two very distinct groups: terrestrials and giants.

Since water, as water vapour and ice, is one of the well identified constituents of interstellar matter, its presence can be expected in the cloud of gas and dust (the protosolar cloud) from which the Sun and its family arose. In this chapter we shall revisit the main events of the history of the formation of the solar system, and see how water played a major part in that saga.

THE SOLAR SYSTEM TODAY

Of what does the modern solar system consist? The Sun, at the centre of the system, is surrounded by nine planets, all orbiting in the same direction; their revolutions are 'direct' (anticlockwise) as viewed from above the Sun's north pole. Most of the planets have almost circular orbits close to the plane of the Earth's orbit (the ecliptic plane). Six of the nine planets rotate on their axes in the same (direct) sense as their revolution about the Sun (Figure 3.1).

Within the solar system, the most convenient yardstick for measuring distances is the Astronomical Unit (AU) – the average distance between the Earth and the Sun, or, more technically, the semi-major axis of the Earth's orbit. With the exception of Pluto (the planet furthest from the Sun), the planets can be classified into two distinct categories. The four terrestrial planets of the inner solar system (Mercury, 0.4 AU from the Sun; Venus, 0.7 AU; Earth, 1 AU; and Mars, 1.5 AU) are relatively small and very dense (between 3.9 and 5.5 g/cm^3). Earth is the largest of them. They have few satellites, or none at all, and their atmospheres constitute a very small fraction of their total mass. Further out from the Sun orbit the four giant planets. They are much more massive than the Earth, but far less dense. Enormous Jupiter, 5 AU from the Sun, is eleven times wider than the Earth and 318 times more massive. Saturn (10 AU from the Sun) is nine times wider than the Earth and 95 times more massive. Uranus and Neptune (19 and 30 AU from the Sun) are respectively fifteen and seventeen times more massive than Earth, and their diameters are both about four times greater than Earth's. These four planets are low-density bodies. The figures for Jupiter and

---

A photo-montage of the solar system. The sizes of the planets and their distances from the Sun are not shown to scale.

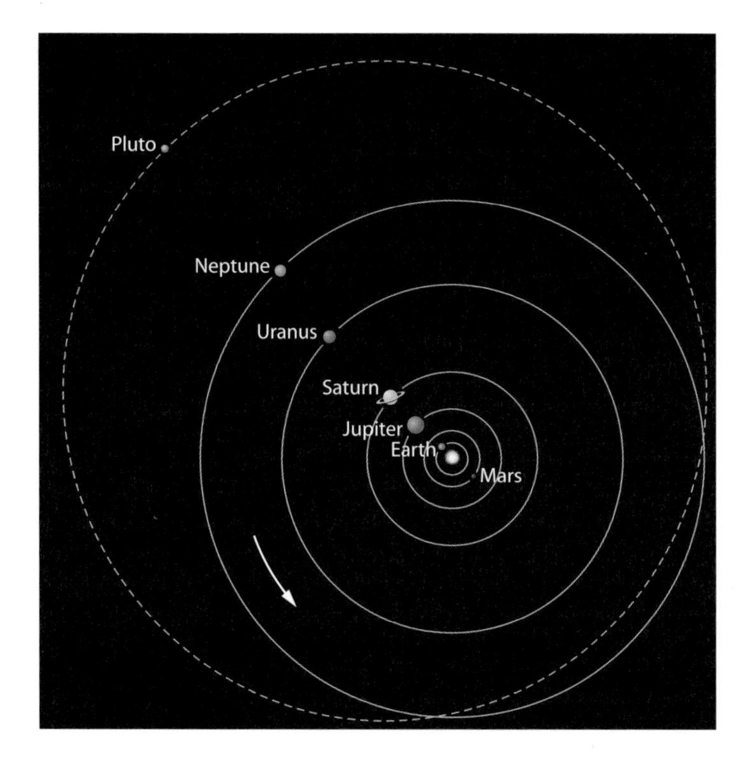

Figure 3.1. The solar system. Most of the planets move in near-circular orbits close to the plane of the Earth's orbit (the ecliptic plane). Of the nine planets, six spin in the same direction as that of their revolution around the Sun. The orbits of Mercury and Venus are not shown.

Saturn are, respectively, 1.3 and 0.7 g/cm^3, and for Uranus and Neptune, 1.2 and 1.7 g/cm^3. The giants have large numbers of satellites, and all four have ring systems which vary considerably in nature from planet to planet. The spectacular rings of Saturn have been observed for centuries, but those of the other giants are much more tenuous and were discovered only in recent decades (see fact box, p. 56).

Between the realms of the terrestrial and the giant planets lies the main asteroid belt. The asteroids are smaller, solid bodies made mostly of rocky, metallic and carbonaceous matter. Their diameters vary from a few tens of kilometres up to about a 1,100 km for the largest, Ceres. Like Ceres, Pallas and Vesta were discovered early in the nineteenth century. Their diameters exceed 500 km, while all the other asteroids are smaller. The main asteroid belt lies between 2 and 3.3 AU from the Sun. Its members' orbits are not all concentrated along the ecliptic, but exhibit a certain inclination to it, usually of about 10°. Other families of asteroids have strongly elliptical orbits, and some may approach relatively close to the Earth. These are known as 'Earth-grazers'. In the past, asteroids have collided with our planet. It seems that such an event may have caused the extinction of many Earth species, including dinosaurs 65 million years ago.

Asteroids are not the only minor bodies wandering the solar system. Since the early 1990s a new and very distant population of objects has been revealed: the trans-Neptunian bodies of the Kuiper Belt, orbiting between 30 and 100 AU from

Figure 3.2. Jupiter and the four Galilean satellites. Callisto is in the foreground, Ganymede is at lower left, Europa is at the centre, and Io, the closest of the four satellites to Jupiter, is shown to the far left of the planet. This is a montage of Voyager 1 photographs taken at the time of the Jupiter encounter in March 1979.

Figure 3.3. Saturn and its largest satellites. In the foreground is Dione, to its right are Mimas and Tethys, while Enceladus and Rhea are at upper left and Titan is at upper right. This is a montage of Voyager 1 photographs taken as the probe encountered Saturn in November 1980.

Figure 3.4. A false-colour image of Saturn's rings. Image taken by Voyager 2 in August 1981, from a distance of about 9 million km.

the Sun. The Kuiper Belt's most famous member is Pluto, discovered in 1930, and until recently considered a planet in its own right. Like most of the Kuiper Belt objects, Pluto – a small body only 2,300 km in diameter – is composed mostly of ice. Its orbit is highly inclined and elliptical, suggesting that it is very different from the two categories of planet described above. Telescopes and other instruments of ever-increasing power add new objects to the list of Kuiper Belt objects almost daily, and more than 1,000 are now known.

Finally, we come to another class of small objects, known since ancient times and the stuff of legend: the comets. As we saw in Chapter 2, comets are very small bodies (usually a few kilometres in diameter), made mostly of ice and orbiting the Sun in very eccentric paths. They therefore spend most of their time at great distances from the Sun, in a cold void where no thermal modification is possible and the risk of collisions is low. This is why astronomers are so interested in them: they represent material as it was when the solar system was very young, and provide clues to the conditions and physico-chemical processes of that era. Periodically, comets approach the centre of the solar system and pass near the Sun and the Earth. Warming up, the ice on their surfaces sublimates, and gas and dust are thrown off to form the coma, which reflects the light of the Sun and may offer a spectacular sight in the sky. This phenomenon – unexplained throughout the ages – led to the terrors and superstitions associated with bright comets in

The planets of the solar system

The planets of the solar system can be divided into two distinct categories: the terrestrial group, within 2 AU from the Sun, and the giants, beyond 5 AU. The terrestrials (Mercury, Venus, the Earth and Mars) are rocky planets, small in size and very dense. They have few satellites. With the exception of Mercury they have stable atmospheres. The giant planets are massive, but not very dense. They all have ring systems and large numbers of satellites. The two biggest – Jupiter and Saturn, the giants nearest to the Sun – are mostly gaseous, while the other two – Uranus and Neptune – are essentially icy in nature. Beyond Neptune the last planet, Pluto, is now thought to be one of the largest elements of a new class of recently discovered objects: the trans-Neptunian or Kuiper Belt objects. Table 3.1 lists the orbital and physical characteristics of the planets of the solar system.

Table 3.1. Physical and orbital characteristics of the planets of the solar system.

Planet	Mean distance from Sun (AU)	Orbital period (years)	Mass (M_E)	Density (g/cm^3)	Rotation period	Atmospheric composition
Mercury	0·39	0·24	0·06	5·43	58·6d	–
Venus	0·72	0·61	0·81	5·20	243·0d	CO_2, N_2
Earth	1·00	1·00	1·00	5·52	23·9h	N_2, O_2
Mars	1·52	1·88	0·11	3·93	24·6h	CO_2, N_2
Jupiter	5·20	11·86	317·9	1·33	9·9h	H_2, He
Saturn	9·54	29·42	95·16	0·69	10·7h	H_2, He
Uranus	19·19	83·75	14·53	1·32	17·2h	H_2, He
Neptune	30·07	163·72	17·14	1·64	16·1h	H_2, He
Pluto	39·48	248·02	0·002	1·94	6·4d	N_2

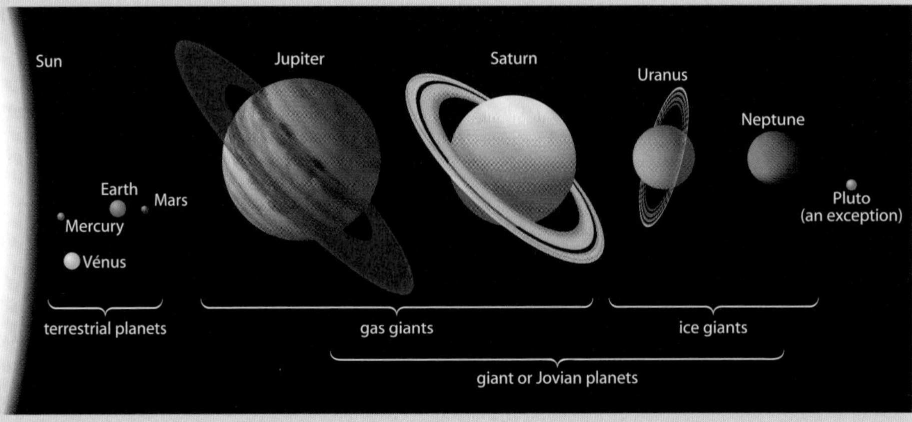

Figure 3.5. Relative sizes of the planets

history. However, early in the eighteenth century, Edmond Halley began to unravel the mystery of comets and their origins when he analysed their paths in former times.

Before we leave this brief description of the solar system, one last question remains. How old is it? We know the answer to this question to a remarkable degree of accuracy, thanks to analyses of isotopic ratios in sample material from Earth, the Moon and meteorites. Meteorites are stones from space, originating within asteroids of various families. Dating them relies on the study of long-lived isotopes (see fact box, p. 57). All measurements tend towards a single value: our Earth, its Moon and the meteorites were formed 4.55 billion years ago. As shown

Determining the age of the solar system through isotopic dating

Most chemical elements have one or more isotopes; that is, there are one or more atoms with the same number of protons and electrons (and therefore the same charge) as the element in question, but a different number of neutrons (and therefore a different atomic mass). Some of these isotopes are not stable, but radioactive, and disintegrate over time into another element of the same atomic mass, according to an exponential law. The rate of disintegration is estimated by measuring the period of the isotope, the time at the end of which the number of radioactive isotopes is divided by a factor *e* (2.7). In some cases these periods are very brief on the cosmological scale; for example, aluminium-26 (^{26}Al) is transformed into magnesium-26 (^{26}Mg) in a period of 700,000 years. Other elements have very long periods – longer even than the current age of the Universe. For example, potassium-40 (^{40}K) disintegrates into argon-40 (^{40}A) over 70 billion years, and uranium-238 (^{238}U) is transformed into lead-238 (^{238}Pb) over 66 billion years. The disintegration of so-called 'parent' radioactive nuclei leads to an excess of 'daughter' nuclei of the same atomic mass, which may be measured by comparing the abundance with that of other (stable) isotopes of the 'daughter' element. For example, if we use the long-period pair rubidium-87 and strontium-87 (^{87}Rb, ^{87}Sr), we compare the abundances of radioactive elements with those of strontium-86 (^{86}Sr), which is a stable element and therefore an indicator of the initial abundance of the strontium. Thus, the date at which the radioactive element was incorporated into the sample in question can be determined. This process is known as isotopic dating.

This method has been used on three types of object: terrestrial rocks, lunar samples and meteorites. To measure the age of the solar system we use long-period 'clocks' such as (^{40}A, ^{40}K), (^{87}Rb, ^{87}Sr) and (^{238}U, ^{238}Pb). One fundamental result has been the determination of the ages of Earth material (from 3.8 to 4.3 billion years), lunar material (4.4 billion years) and the parent bodies of the most ancient meteorites (4.56 billion years). This last value indicates the age of the solar system.

by the model described below, the Sun must have formed somewhat earlier – perhaps about 100 million years before. This provides us with a reliable indicator of the age of the Sun, which is now about halfway through its life.

The bodies comprising the solar system therefore exhibit a great variety of physical and chemical characteristics, and are of very different sizes, densities, temperatures, chemical composition, orbital parameters, and so on. All these elements must be taken into consideration when we seek to construct a global model of the formation of the solar system. This is an ambitious project, and much remains to be known; but today's astronomers may have cause for satisfaction, since the accepted model (which we shall call the Mizuno–Pollack model) is largely in accordance with the diverse characteristics observed in the various types of object which inhabit the solar system. This is the scenario which will now be described.

THE COLLAPSE OF THE PROTOSOLAR CLOUD

Let us begin with a simple statement. The Sun is a very ordinary star – just one of 10 billion of its kind within our Galaxy. It makes sense, while trying to piece together the story of its birth, to observe other stars in the process of formation and the regions in which they are formed. These regions are dense, rapidly rotating molecular clouds, contracting until they settle into a disc. From its centre, the nascent stars will emerge. How have we been able to observe these early stages in the evolution of stars? Over the last few decades, much light has been thrown onto this research by developments in the fields of infrared and millimetre astronomy. More recently still, new techniques have improved the quality of images and have revealed celestial bodies in ever greater detail. All these advances have enabled us to determine the physico-chemical parameters (pressure, temperature, molecular abundances) of star-forming clouds, as have new images showing protoplanetary discs surrounding some nearby stars. We have been able to visualise the first phases in the birth of stars due to instruments onboard the Hubble Space Telescope (working in the visible domain) and the ISO (infrared imaging and spectra), in addition to the charts created by millimetre-region radio telescopes. This whole array of detectors is backed up by digital simulations of evolution within protostellar clouds and their collapse into discs.

Independently of our observations of protostellar discs in our vicinity, we have good reason to believe that the solar system was formed in a similar way. Strong evidence is provided by its current configuration: the near-circular orbits of the planets, both terrestrials and giants, lie very close to the ecliptic plane; all the planets move in the same direction, 'direct' as viewed from above the Sun's north pole; and most of them spin in the same sense. All these properties fit naturally into a scenario of planetary formation within a rotating disc of protosolar matter. All these facts – underpinned by the theory of universal gravitation formulated by Isaac Newton in the late seventeenth century – were

Figure 3.6. Star-birth regions in the Eagle Nebula (M16). This image was taken in 1995 with the Hubble Space Telescope's wide-field planetary camera (WFPC2). The 'pillars' are relatively cool columns of molecular hydrogen and dust linked to the inner wall of a dense interstellar cloud.

incorporated into the visionary thinking of Immanuel Kant (1755) and Pierre-Simon de Laplace (1809), who put forward the first versions of the nebular theory now universally accepted.

FROM PROTOPLANETARY DISC TO PLANETESIMALS

The protosolar nebula took very little time (certainly of the order of a million years) to settle, compared with the lifetime of the solar system. In a few million years, the central material of the disc became concentrated into a nascent star – its surface temperature being certainly around a few thousand K. Its core temperature, on the other hand, reached a few million degrees, igniting the nucleosynthesis which has progressively transformed lighter elements (hydrogen,

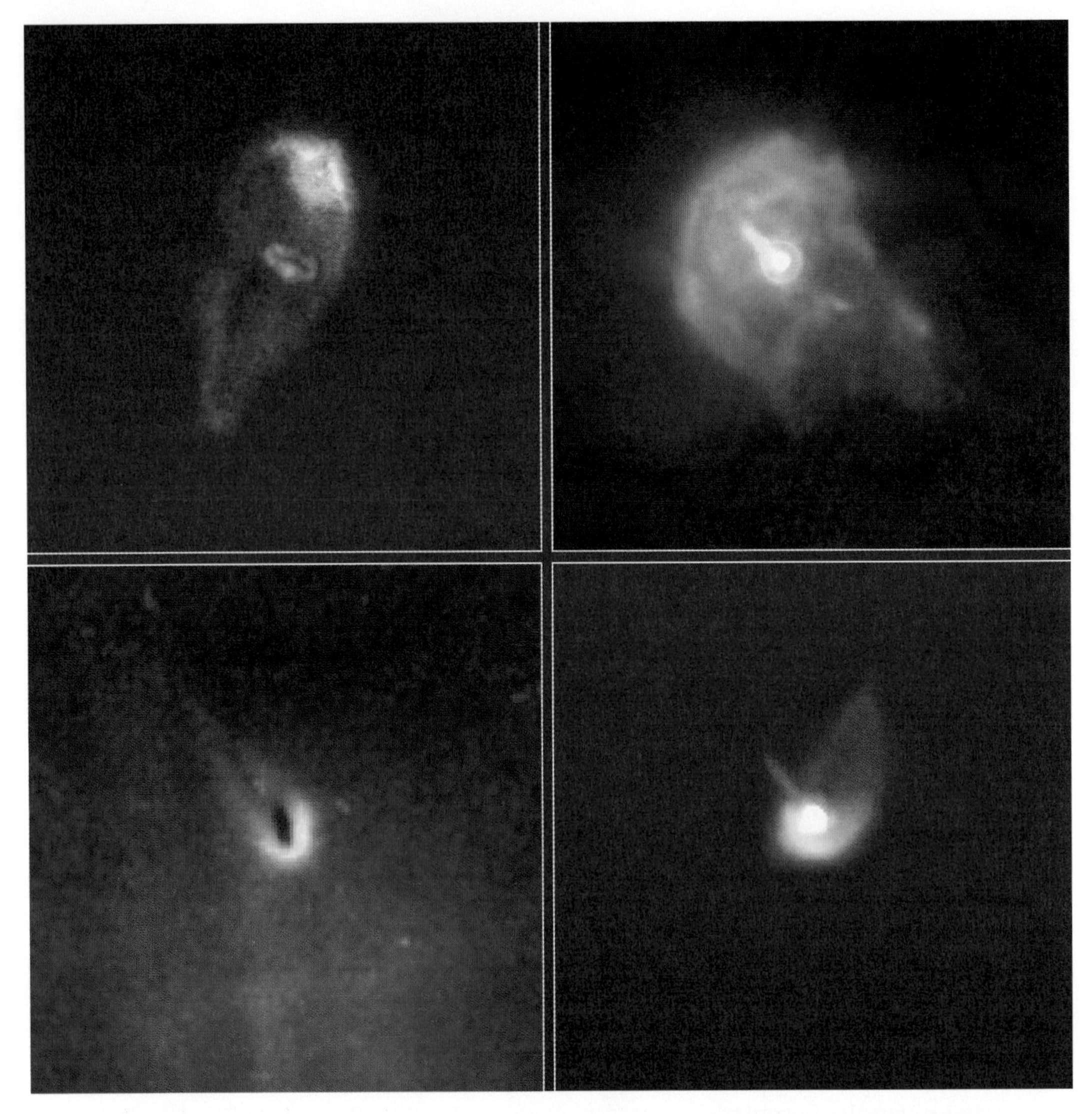

Figure 3.7. Protoplanetary discs in the Orion Nebula. Images taken with the Hubble Space Telescope's wide-field planetary camera (WFPC2).

deuterium, and so on) into heavier ones (helium, and later, carbon, nitrogen and oxygen). In this early stage of its life, the Sun underwent a very active T Tauri phase (named after a young star in which this behaviour has been observed). Rotating much faster than it does today, and possessing a much more intense magnetic field, the juvenile Sun blasted out, along its axis of rotation, a jet of energetic particles, much stronger than its present solar wind. This intense stream – a flux of energised particles continuously emitted by the Sun – was to play an important part in the history of planetary formation.

Figure 3.8. A void created by a planet in a disc of gas. (Courtesy G. Bryden.)

What remained of the protosolar nebula (between 10% and 1% – the rest having been incorporated into the Sun) settled into the protoplanetary disc, extending out to a distance of several tens of AU. Naturally, temperature and density decreased from the centre outwards. In the vicinity of the Sun, the temperature must have been more than 2,000 K, while at a distance of 30–50 AU it was probably less than 100 K. The temperature fell gradually as time passed and the disc cooled.

What was the composition of the protoplanetary disc? Certainly there was gas, but with it were dust, metals, rock and ice – their abundances similar to those in the cosmos in general. Hydrogen was the most abundant element, and, since its condensation temperature is very low (just a few K), molecular hydrogen was present in gaseous form. The same was true of helium and the other noble gases – neon, argon, krypton and xenon. In line with the table of cosmic abundances (Chapter 1), the most common elements after hydrogen and helium were carbon, nitrogen and oxygen. In a hydrogen-dominated environment they associated to become CH_4, NH_3 and H_2O, together with CO and, in lesser quantities, CO_2. In what state, solid or gaseous, did these molecules exist? This depended upon their distance from the Sun and their position on either side of the 'ice line'. This line – defined by the condensation temperature of water (the triple point is at approximately 273 K), and therefore at a certain distance from the Sun – marked the frontier between the gaseous and solid states of the molecules, respectively, to the Sunward side of it and beyond it. When the planets were formed, the ice line was situated at about 4.5 AU from the Sun. The slow cooling of the nebula has brought the present position of the ice line to around 2 AU from the Sun.

The role of water was important for two reasons; first, given the cosmic abundances of H and O, the H_2O molecule was the most common among the ices; second, compared with other ices (methane, ammonia, carbon dioxide 'snow', and so on), water ice was the most refractory; that is, it sublimated at a higher temperature (Figure 3.9). It was therefore water that determined the position of the ice line, beyond which all protosolar material (with, of course, the exception of hydrogen and helium) was in solid form.

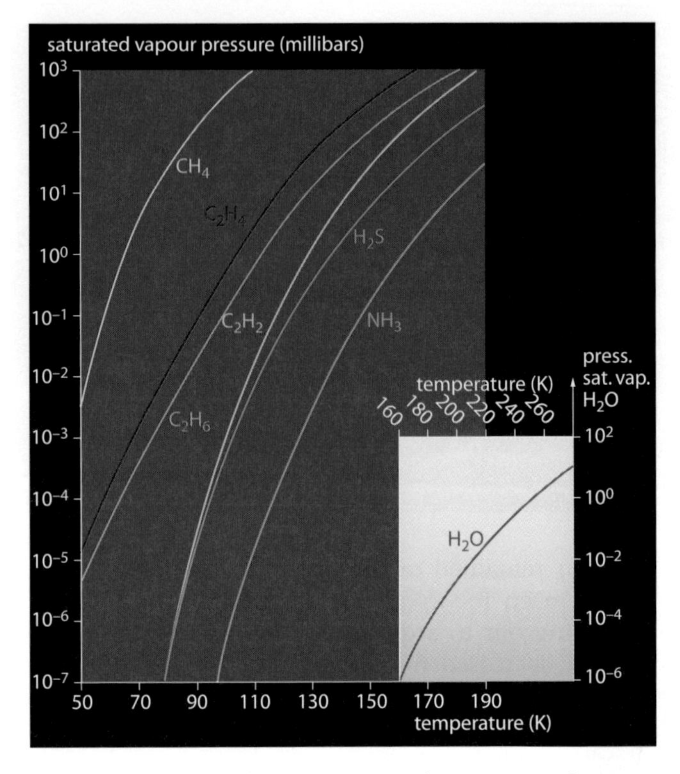

Figure 3.9. Saturated vapour pressure/temperature diagrams for condensable gases in outer planets. The insert shows the saturated vapour pressure for water as a function of temperature (scale at top). The curves show that water vapour is the gas which, at a given pressure, condenses at the highest temperature. (From S.K. Atreya, *Atmospheres and Ionospheres of the Outer Planets and Their Satellites*, Springer-Verlag, 1986.)

TERRESTRIAL PLANETS AND GIANT PLANETS

The ice line played a major part in the separation of the terrestrial and the giant planets. The first embryonic planets formed from multiple collisions between solid particles in the protoplanetary disc. At this stage, gas played little or no part in the proceedings, The solid particles, by way of macromolecules, underwent successive collisions to form micron-sized, millimetre-sized and then centimetre-sized conglomerates – and the process of *accretion* was under way. Imagine a chunk of snow rolling down a slope as part of an avalanche, soon growing as it picks up new snow along the way (Figure 3.10).

The embryonic planets thus grew, and when their diameters reached a few tens of kilometres their gravitational fields began to assert themselves. The first effect was that the planets became spherical; and the second was the amplification of the accretion process. The larger the embryos grew, the more surrounding material they drew in, and their eventual size depended upon the amount of available solid material in the protoplanetary disc. The ice line was of crucial importance at this time.

On the Sunward side of the ice line, at distances less than about 3 AU from the Sun, the only elements existing in solid form were the heavier ones, especially metals and silicates. However, the latter were in relatively short supply,

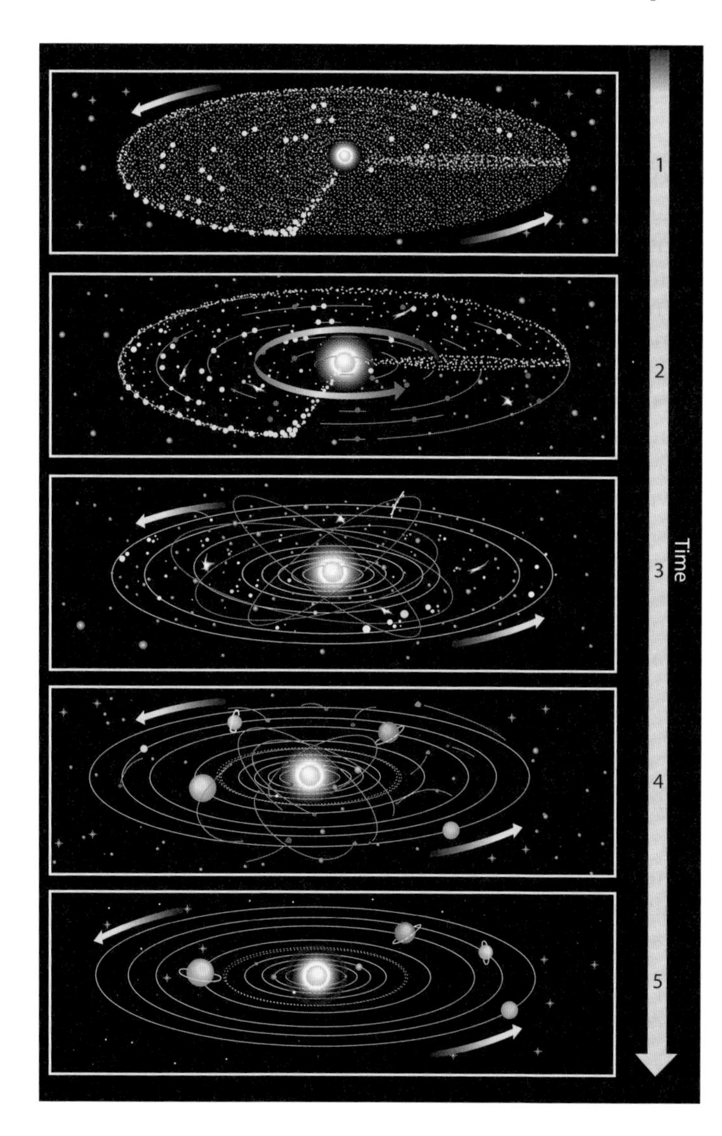

Figure 3.10. Formation of the solar system according to the theory of progressive accretion. (1) Protosolar disc (gas, droplets and dust). (2) Formation of 'pebbles' and aggregates (between 1 cm and 1 m across). (3/4) Objects increase in size (planetesimals) to between 1 m and 1 km as a result of repeated collisions; each has its own orbit, and planets form from the stock of planetesimals. (5) Concentration of planets and planetesimals in the plane of the ecliptic. The ice line now lies between the orbits of Mars and Jupiter. (From C. Allègre, 1985.)

as can be seen from the distribution of cosmic abundances (Figure 1.5). The amount of solid material was therefore limited. Numerical models show that, given an initial mass for the disc of about 0.1 that of the Sun, about ten embryonic planets between the sizes of Mars or the Moon could be formed, after which the number of embryos would be reduced to leave just a few planets – the terrestrials, of which the Earth is the most massive.

Beyond the ice line, the amount of available solid material within the disc was much greater than that of the inner solar system, due to the presence of ices. In fact, still in keeping with the distribution of cosmic abundances (Figure 1.5), the contribution of ice to the whole mass was far more important than that of

The sequence of condensation

When the protosolar nebula collapsed into a disc, its core temperature decreased as a function of distance from the Sun. The whole disc cooled as time went on. The different constituents of the disc condensed progressively, the first to condense being the most refractory. Close in to the Sun, metals condensed first, and further out, rocks. This phase occurred more than 4 billion years ago. In the case of the refractory elements, the order of condensation was Al, Ti, Ca, Mg, Si, Fe, Na and S. Most of the stable types condensed at temperatures between 700 and 2,000 K. It is observed that the expected abundances of the main refractory elements are in very good agreement with the abundances of refractory components measured within primitive meteorites (C3 chondrites), for example, the Allende meteorite.

Beyond the boundary known as the ice line, ices condense at temperatures below 200–300 K: at first H_2O, then NH_3, CO_2, CH_4, and so on. Bearing in mind the much greater relative abundance of H, C and O, the existence of these ices has led to the formation, from planetesimals, of objects more massive than the terrestrial planets, with gravitational fields capable of drawing in the surrounding nebular material to create gas giants.

The chemical composition of planetary atmospheres

In the light of the scenario of accretion, we can predict the chemical composition of the atmospheres of both the terrestrial planets and the giant planets. In thermochemical equilibrium, the carbon and nitrogen components participate in the following reactions:

$$CH_4 + H_2O \rightleftarrows CO + 3H_2$$
$$2NH_3 \rightleftarrows N_2 + 3H_2$$

These reactions proceed from left to right at high temperature and low pressure, and from right to left at low temperature. Another reaction involved is:

$$CO + H_2O \rightleftarrows CO_2 + H_2$$

Due to conditions of thermochemical equilibrium, large numbers of collisions will occur between molecules, the environment being of considerable density. Such conditions are observed in the sub-nebulae of Jupiter and Saturn, where the environment is at once cold and dense. Here, carbon and nitrogen are found principally as components in methane and ammonia, in agreement with photochemical models. Nearer to the Sun, however, where higher temperatures prevail, no sub-nebulae are found around the terrestrial planets. Here, CO and N_2 dominate, CO reacts in its turn with water to form CO_2, and hydrogen, too light to be held by the terrestrials' gravity, escapes into the interplanetary medium. Thus can be explained the chemical composition of the terrestrial planets, where CO and N_2 are preponderant (together with H_2O, in the case of the Earth).

Our thermochemical model therefore effectively explains the global chemical composition of the planetary atmospheres. However, questions arise if we look at the atmospheric composition of the satellites in the outer solar system, and of comets. For example, the atmosphere of Titan, Saturn's largest satellite, is mostly made of N_2 and CH_4, apparently contradicting the model as described. It is possible that Titan's interior, formed within the planet's sub-nebula, consisted of H_2O, NH_3 and CH_4, all in the solid state. The ammonia and methane would have been outgassed from the globe as the satellite formed, and NH_3 would have produced, through a photochemical reaction, the N_2 now observed in Titan's atmosphere.

As for comets, they were probably formed from planetesimals which themselves formed in unstable conditions, in a rarefied environment where collisions were unlikely. Ammonia and nitrogen are very uncommon in comets. Molecular nitrogen – which is not very reactive, and spectroscopically inactive – is not easy to detect in comets, while CO appears to be the most abundant cometary molecule after water.

silicates and metals. The formation of large, icy planetary nuclei was therefore possible in the outer solar system, their masses attaining 10–15 times that of the Earth. Their gravitational fields became strong enough to draw in the protosolar gas from their surroundings. Just as the primordial solar nebula had settled into a protoplanetary disc because of its own gravity, the future giant planets, like miniature solar systems, were surrounded by sub-nebulae, within which accretion created rings and some of the planets' satellites. This was not the process of formation of the terrestrial planets, as their gravity was not strong enough to capture the surrounding gas - which, it will be recalled, was principally hydrogen, the lightest of all the elements.

Here we see the reason why the terrestrial planets are so different from the giants. They are relatively small and dense - a result of the heavy elements in their cores. They have either no satellites or very few, and lack rings, having never undergone the sub-nebular stage. Their atmospheres have not been drawn down from protosolar gases, but are the products of outgassing (especially volcanism) and the meteoritic impacts which occurred so frequently during the first billion years of the history of the solar system.

A BRIEF CHRONOLOGY OF EVENTS

The giant planets

After a few million years, the giant planets had formed. The first were Jupiter and Saturn. Jupiter was in prime position, situated as it was just beyond the ice line and therefore able to draw large numbers of planetesimals into its path. This is doubtless the reason why it is considerably larger than the other giant planets. In the case of Saturn, a little further out, the same process occurred, but to a lesser degree. Over a few million years, Jupiter and Saturn had collected most of their material. Then came the Sun's T Tauri phase, and gas and smaller planetesimals were swept outwards. Some of the hydrogen was swept away by the solar wind, and little protosolar gas remained when Uranus and Neptune reached the critical size of 10–15 Earth masses beyond which accretion could take place. This is why these two planets are much smaller than Jupiter and Saturn. With masses, respectively, of 15 and 17 Earths, they were able to collect relatively little of the protosolar gas, and largely consist of icy nuclei. Indeed, they are sometimes referred to as 'ice giants', while Jupiter (318 Earth masses) and Saturn (95 Earth masses), composed mostly of hydrogen and helium, are the 'gas giants'.

Asteroids and comets

The formation of the asteroid belt (Figure 3.11) was no doubt contemporaneous with the formation of Jupiter. Indeed, the very proximity of this supermassive body must have hindered the formation of a planet at 3 AU from the Sun, as the tidal forces of the jovian gravity field destroyed the largest embryonic planets. It seems, then, that the asteroid belt represents the remains of an aborted planet. Its total mass is less than 0.001 that of the Earth.

Comets, however, must have formed at the same time as Uranus, Neptune and the Kuiper Belt. The initial reservoir of comets later ejected into the Oort Cloud lay in the zone of the ice giants, while other comets formed within the Kuiper Belt. It is therefore reasonable to suppose that the comets were born about 10 million years after the appearance of the protoplanetary disc. Together with asteroids and the giant planets, comets are among the most primitive objects in the solar system.

The 'regular' satellites of the giant planets

The 'regular' satellites and ring systems of the giant planets are also primitive objects. Satellites are known as regular when they orbit in the equatorial plane of their parent planet, on near-circular paths. Among these are the Galilean moons of Jupiter, Saturn's satellite Titan, and the large icy satellites orbiting Saturn and Uranus. Other satellites, with inclined and elliptical orbits, are the result of captures. The regular satellites were formed within the disc created when sub-nebular material settled around the giants. The process therefore resembled that

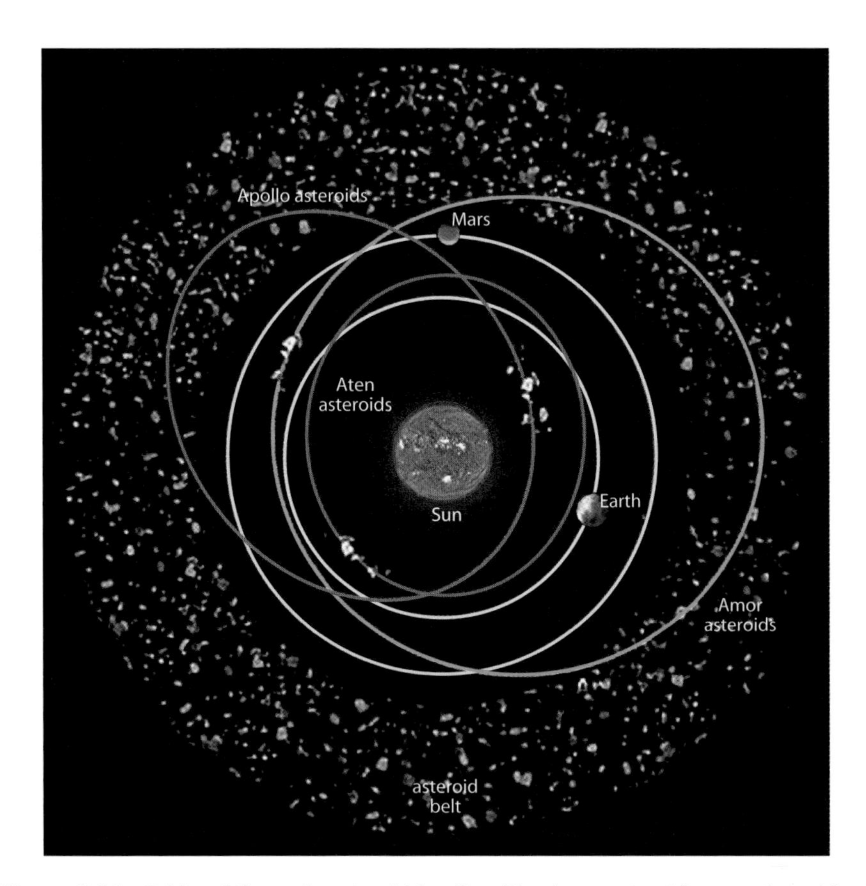

Figure 3.11. Orbits of the main asteroid families. The Amor asteroids cross the orbit of Mars, and the Apollo asteroids cross the orbit of Earth. The Aten group ateroids, by contrast, spend most of their time within Earth's orbit, crossing it rapidly to be lost in the Sun's glare.

of planetary formation at the heart of the protoplanetary disc, but on a much reduced scale which speeded the process. Confirmation comes from the Uranian system and its peculiar configuration. Uranus has a distinctive feature: its axis of rotation is quite close to the plane of the ecliptic, while the axes of the other planets are all more or less perpendicular to it (with axial inclinations within around 20°). Yet all of Uranus' regular satellites orbit in its equatorial plane (perpendicular to the ecliptic). There is a possible simple explanation for this phenomenon. A very massive embryonic planet might once have collided with Uranus, resulting in a considerable shift in its axial tilt but not in its being shattered. The debris from the collision must have augmented the disc, and contributed to building the regular satellites. Since the principal Uranian satellites travel in the planet's equatorial plane, the collision must have happened early in Uranus' history, before its sub-nebular material settled into a disc.

Figure 3.12. Halley's comet. Like asteroids and giant planets, comets are among the most primitive objects in the solar system.

The terrestrial planets

The terrestrial planets were probably formed at a later time. The accretion of the embryos must have taken longer because, on this side of the ice line, less material would be available. If the numerical models are to be believed, these planets took between 10 million and 100 million years to form – the rate of accretion decelerating during the last stages. Then came a period of intense bombardment. In this particularly violent phase, the rate of meteoric infall was very high – and the most ancient craters on the surfaces of the Moon, Mars and Mercury bear witness to this (Figure 3.13). It was certainly during this era that the greater part of the atmospheres of the terrestrial planets was delivered by meteorites (with the exception of Mercury, which lacks the gravitational pull to retain an atmosphere), the rest being due to outgassing from within. This turbulent period ended after about a billion years.

Figure 3.13. A crater on the Moon. Lunar craters, like those on Mercury and Mars, bear witness to the period of intense meteoritic bombardment which marked the first billion years of the solar system's history.

WHERE DO WE LOOK FOR WATER IN THE SOLAR SYSTEM?

Having looked at the main types of object in the solar system, and main traits of their history, we can now begin to identify those where water may be present.

Since material accreted was mostly in the form of solid particles, and water was present in solid form only beyond the ice line, then it is in the outer solar system that we must begin our search for water.

The outer solar system

Comets head the list of candidates. As Fred Whipple predicted in the 1950s, comets are about 80% water. Their ice begins to sublimate as the nucleus approaches the Sun, at a distance of about 2–3 AU. Thus, water vapour could be observed in the coma of comet Halley at perihelion passage in 1986, and in comet Hale–Bopp's coma when it was 3 AU distant. It was also possible to detect water ice in the vicinity of the nucleus (see Chapter 4).

Water vapour is also present in the giant planets. It was, naturally, incorporated into their initial nuclei along with the ice. After the phase of

Chronology of the formation of the solar system

The chronology below (after M. Garlick, 2002) is an approximation of events leading to the birth of the objects of the solar system, according to the protosolar-nebula model now currently favoured. The epochs given are of course only estimates based on today's knowledge, and are of necessity not precise. There remain considerable uncertainties about the time it took planets to form, and the mechanism of the accretion of large bodies from planetesimals is still not very well understood.

t (millions of years)	Event
0	Existence of giant molecular cloud
2	Collapse of cloud and formation of dark solar globule
2.13	Emergence of rotating proto-Sun
2.2	Formation of planetesimals and protoplanets
2–3	Formation of giant gas planets (Jupiter, Saturn) and asteroids
3–10	T Tauri phase of the Sun (intense activity, strong bipolar solar wind)
3–10	Formation of icy giant planets (Uranus, Neptune) and comets
3–10	Formation of regular (non-captured) satellites of outer solar system
30–50	Sun moves onto Main Sequence (hydrogen into helium)
10–100	Formation of terrestrial planets
100–1,300	Massive bombardment (planets and satellites collide with protosolar disc debris)
700–1,300	Formation of atmospheres of terrestrial planets (outgassing and meteoritic impacts)
4,500	Formation of (short-lived) planetary rings

gravitational collapse of the surrounding gaseous material, and the resultant internal heating, water and other components were incorporated into the deep atmosphere. Observations of Jupiter and Saturn confirm this. However, ISO observations of the four giant planets led to a surprise: water vapour was detected not only deep down within them, but also in their stratospheres, and in the atmosphere of Titan. It will be seen in Chapter 5 that this indicates a permanent interplanetary oxygen flux, possibly of micrometeoric origin, captured by the gravitational field of the giants. Other impacts, on a much larger scale, are known to occur: for example, when comet Shoemaker–Levy 9 collided with Jupiter in July 1994. This event caused the formation of water in Jupiter's stratosphere – water that could still be observed several years later.

As for water ice, it has been found in many locations in the outer solar system. It is the main constituent of Saturn's rings, of its satellites without atmospheres, and of three of Jupiter's Galilean satellites: Europa, Ganymede and Callisto.

An indirect proof of progressive accretion: abundances of the elements

The progressive accretion model seems plausible, but what proof have we of its validity? For several decades, another model was set against it: that of the massive disc, within which the giant planets formed by gravitational contraction, because of local instabilities due to inhomogeneities in density. The accretion model (now known as the standard model) has asserted itself over the last two decades due to a new test favouring it: the measurement of the relative abundances of the elements of which the giant planets are constituted.

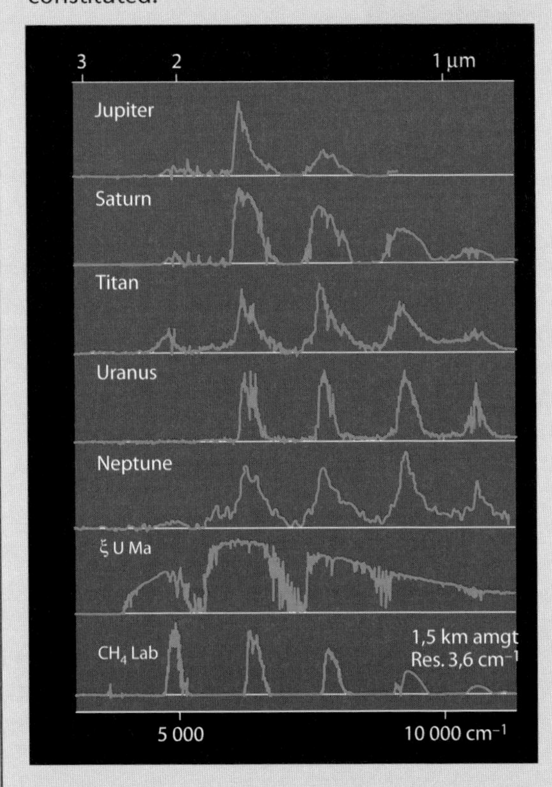

Figure 3.14. The spectra of the giant planets in the near-infrared. These measurements are the result of ground-based high-resolution spectroscopy. The spectrum of the star ξ Ursae Majoris shows those spectral domains where absorption by the Earth's atmosphere predominates. It can be seen that the planetary spectra exhibit strong methane absorption. The methane spectrum is shown at the bottom of the figure. (From H. Larson, *Annual Rev. Astron. Astrophys.*, **18**, 43, 1980.)

To return to the 'massive disc' scenario: giant planets are said to form directly out of the local protosolar nebula, in which case the heavy elements (heavier than helium) in them will be in the same proportions as in the nebular material (2%), in accord with cosmic abundances. However, in the accretion model these planets form from a solid nucleus of 10–15 Earth masses, surrounded by protosolar material. Within the material, the fraction by mass of the heavy elements is the same as in the nebula (2%), while within the planets it is 100%. The relative abundance of heavy elements compared with hydrogen in giants planets is therefore higher in the accretion model than in the massive disc model. This excess can be easily quantified from the total mass of each planet, assuming an homogeneous mix of heavy elements in the atmosphere following the phase of the collapse of the surrounding nebular material onto the

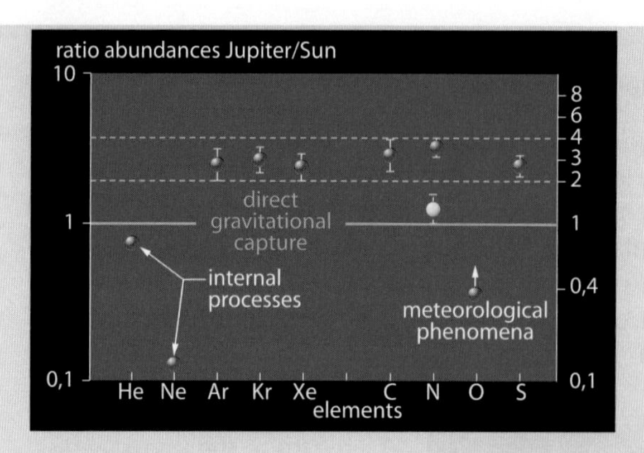

Figure 3.15. Abundance of elements in Jupiter, compared with hydrogen, and normalised with respect to solar values. With the exception of He, Ne and O (doubtless attenuated by internal or dynamical processes), all the elements show an excess of three times the solar value – the cosmic or protosolar values. This is a very telling indication that the formation of the giant planets involved the accretion of an icy nucleus, followed by the infall of the surrounding protosolar gas. (From T. Owen *et al.*, *Nature, 402*, 269, 1999.)

nucleus. Using this hypothesis it is calculated that the factor of this excess for Jupiter is 3, for Saturn 7, and for Uranus and Neptune 30–50. Note that for a planet with a nucleus composed entirely of ice, the excess factor is 50, since the relative amount of heavy elements is 2%.

Even in the 1970s the first measurements of the abundance of methane in the atmospheres of Uranus and Neptune indicated a strong excess in the C/H ratio – a result favouring the accretion model.

In the early 1980s, Voyager measurements showed the C:H ratio for Jupiter enriched by a factor of 3, while the value for Saturn was about 6. Then, the Galileo probe, plunging into Jupiter's atmosphere in December 1995, showed that most of the heavy elements were in excess in relation to hydrogen by a factor of 3. Thus, the 'nucleation' model – proposed by Mizuno in 1980 and later refined by other researchers, notable among them Pollack and collaborators – was definitively confirmed.

It should be borne in mind that there still remain some important questions. The global excess (by a factor of 3) on Jupiter, as measured by Galileo, seems to indicate that all the elements were trapped within the ices of the planetesimals which built Jupiter, including nitrogen and argon, which can be thus confined only at very low temperatures (less than 30 K). This is a temperature far below that of the presumed environment in which Jupiter formed, given its distance from the Sun. Where, then, did these planetesimals come from? The question remains unanswered.

Figure 3.16. The planet Mars, which experienced a modest greenhouse effect.

Europa, above all, has excited much interest among astronomers, since thermochemical models of its interior suggest that, beneath the layer of ice at its surface, there may be liquid water. In decades to come, Europa will doubtless be a priority target for the attention of planetologists and exobiologists! Water ice is also present on the surfaces of more distant objects: for example, Triton, Pluto and its moon Charon, and Kuiper Belt objects. Other ices (N_2, CH_4, CO_2, CO, and so on) are also found, as are products of irradiation which hinder the identification of water (see Chapter 5).

The inner solar system

Asteroids within the main belt contain water, though amounts vary according to their provenance. This is water trapped within planetesimals during their

Figure 3.17. Venus, photographed by the Galileo probe as it flew by the planet in February 1990. The clouds, composed mainly of sulphuric acid, conceal the planet's surface completely.

accretion phase. Laboratory analysis of meteorites helps to determine quantities present, since most meteorites' parent bodies are asteroids. This asteroidal water has been an important factor in our own history, since asteroids are certainly one of the sources of Earth's water.

The terrestrial planets themselves have stored up a considerable amount of water in their atmospheres, as a result of both outgassing (and especially volcanism) and meteoritic and micrometeoritic delivery. The primitive atmospheres of Venus, Mars and the Earth were doubtless very rich in carbon dioxide and water vapour, together with lesser quantities of molecular nitrogen. On Venus, enormous amounts of water vapour and carbon dioxide led to a runaway 'greenhouse effect', responsible for the very high surface temperature on Venus today: 730 K. Venus' water vapour soon disappeared, in a way still not clearly understood. On Earth, due to the existence of liquid water, carbon dioxide was dissolved to form calcium carbonate. A moderate greenhouse effect ensued, and temperatures have since remained globally constant on our planet. Mars - which is smaller and cooler - retains a modest greenhouse effect. Its outgassing has ceased, owing to lack of internal energy, and its falling temperature caused its water - certainly liquid at times during Mars' early history - to become permanent ice, further diminishing the greenhouse effect. Most of its atmosphere seems to have escaped - a process which remains largely unexplained and unquantified. The comparative histories of Venus, Earth and Mars - exhibiting relatively similar conditions early on but later becoming so different from each other - constitute one of the most fascinating problems in modern planetology (see Chapter 7).

4
Comets and water

Composed essentially of water ice, comets have not undergone any gravitationally-induced interior transformation. Their orbits take them far from the Sun to a cold environment where collisions are improbable, so they carry within them vestiges of the early stages in the formation of the solar system. Where do they come from? How did they form? The water of which they are made can offer clues to the answers.

Due to their essential composition (water ice), comets are a natural starting point for our *tour d'horizon*. Their nuclei, containing on average as much as 80% water by mass, also have other ices in much smaller quantities, mixed with siliceous and carbonaceous dust. Because a comet's nucleus is so small - usually just a few kilometres across - it is not differentiated, and has not sustained the internal transformation (due to gravity) typical of more massive bodies such as asteroids, satellites and planets. Its shape is irregular, because it lacks the gravitational force to become spherical. Far from the Sun, in a cold environment with low risk of physico-chemical transformations due to collisions, cometary nuclei carry within them vestiges of the first stages in the formation of the solar system. This is why comets are of such interest to astronomers. They are the most telling witnesses to the dim and distant origins of our solar system.

Since ancient times, comets have inspired fear and superstition. Our ancestors, gazing at the immense, dusty tresses of comets trailing across the sky as they passed near the Sun and the Earth, could have no idea that the bodies responsible for these phenomena were so small - among the smallest in the solar system.

Even after Edmond Halley had thrown light upon the nature of comets in the eighteenth century, they continued to unnerve people, who associated their appearances with dire portents (Figures 4.1 and 4.2).

THE NUCLEUS: A 'DIRTY SNOWBALL'

Two centuries after Halley, a new name came to the fore in the history of cometary physics: the American astronomer Fred Whipple. In a still famous article, published in 1950, Whipple set forth a model of the nature of a cometary nucleus. Since comets formed far from the Sun, they must consist mostly of ice; and because water ice heads the table of cosmic abundances (see Chapter 1), and is the ice which vapourises at the highest temperature, it must be the most abundant form in comets. However, there must also be dust present, since it is involved in the formation of the well-known cometary tails, which, by reflecting sunlight, can produce such an awe-inspiring celestial spectacle. So, in Whipple's now famous phrase: a comet's nucleus is like a 'dirty snowball'.

Several decades passed before Whipple's visionary intuition was confirmed by the direct observation of water in a comet. As we have seen in Chapter 2, nowadays the detection of water molecules is easier in the infrared and

---

Comet Hale–Bopp.

Figure 4.1. An Aztec painting depicting the Emperor Moctezuma observing a comet. The Aztecs saw comets as sinister portents. (From D. Duran, *Historia de las Indias de Nueva España*.)

Figure 4.2. Halley's comet on the Bayeux Tapestry. This section of the tapestry shows King Harold and his advisers, alarmed by the apparition of the comet in 1066. Not long afterwards, Harold would be vanquished by William the Conqueror at the Battle of Hastings.

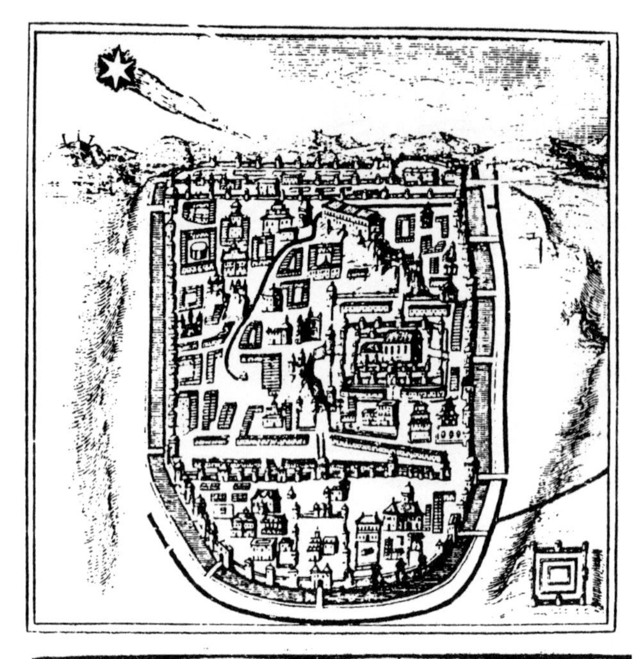

Figure 4.3. The comet of 63 AD over Jerusalem. This was not Halley's comet, which appeared in the year 66. (Courtesy International Halley Watch.)

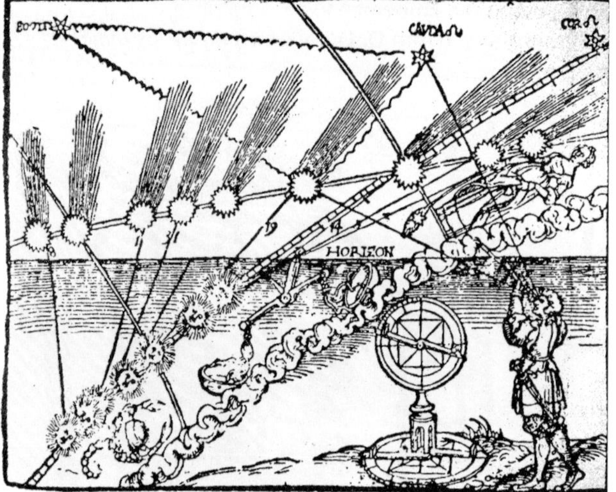

Figure 4.4. Peter Apian (1495–1552), whose observations showed that a comet's tail always points away from the Sun.

submillimetre domains: but before the 1980s, observations of comets were limited to the visible and the ultraviolet (Figure 4.5). Gradually, however, various clues emerged to support the 'dirty snowball' model. Although H_2O was not directly detectable, the hydrogen atom H could be detected (via very intense Lyman-α transitions in the ultraviolet, and Hα transitions in the visible, respectively at 1216 Å and 6563 Å), and so could the OH radical (3090 Å, UV). Also detected was the oxygen molecule O at several different wavelengths in the visible and the ultraviolet. Measurements of the relative observed abundances

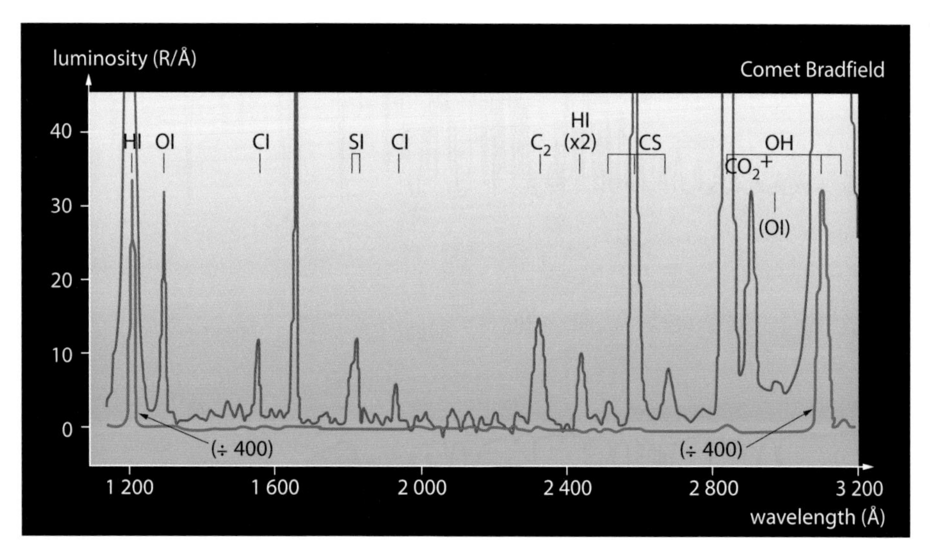

Figure 4.5. The ultraviolet spectrum of comet Bradfield 1979, from the IUE satellite. The signatures of atoms and radicals in the coma are apparent, while the signatures of molecules in the nucleus appear in the infrared. The OH radical produced by the dissociation of H_2O is visible around 3090 Å. (From P. Feldman, *Comets*, 1982.)

showed that the overall H:O ratio was of the order of 2:1, which strongly suggested that H_2O was the dominant component. Other confirmatory results included the observation of the expansion of atomic hydrogen into the outer coma, compatible with photochemical models of the dissociation of H_2O by solar ultraviolet radiation, and the detection of the H_2O^+ ion in the visible domain at 6198 Å, in 1974.

HALLEY'S COMET, 1986: THE FIRST DETECTION OF WATER VAPOUR

The latest return of comet Halley in 1986 provided astronomers with the opportunity to finally detect cometary water vapour. This comet, with a period of 76 years, is, of course, the most famous in the history of cometary physics. It has been known since ancient times, and enabled Halley, who studied observations of its many returns, to unravel some of the mystery surrounding comets. Due to accurate knowledge of its orbit during the 1986 passage, a fleet of five spacecraft (see fact box, p. 83) was sent to investigate it in all sorts of ways. These included imaging the nucleus, spectroscopy of the coma, mass spectroscopy of gas and dust, and analysis of the interaction between the comet and the solar wind. This space spectacular was matched by an unprecedented international observing effort on the ground, involving a large number of telescopes and making best use of the most up-to-date techniques, especially in the infrared and millimetre regions. The investigations of comet Halley in 1986, bringing

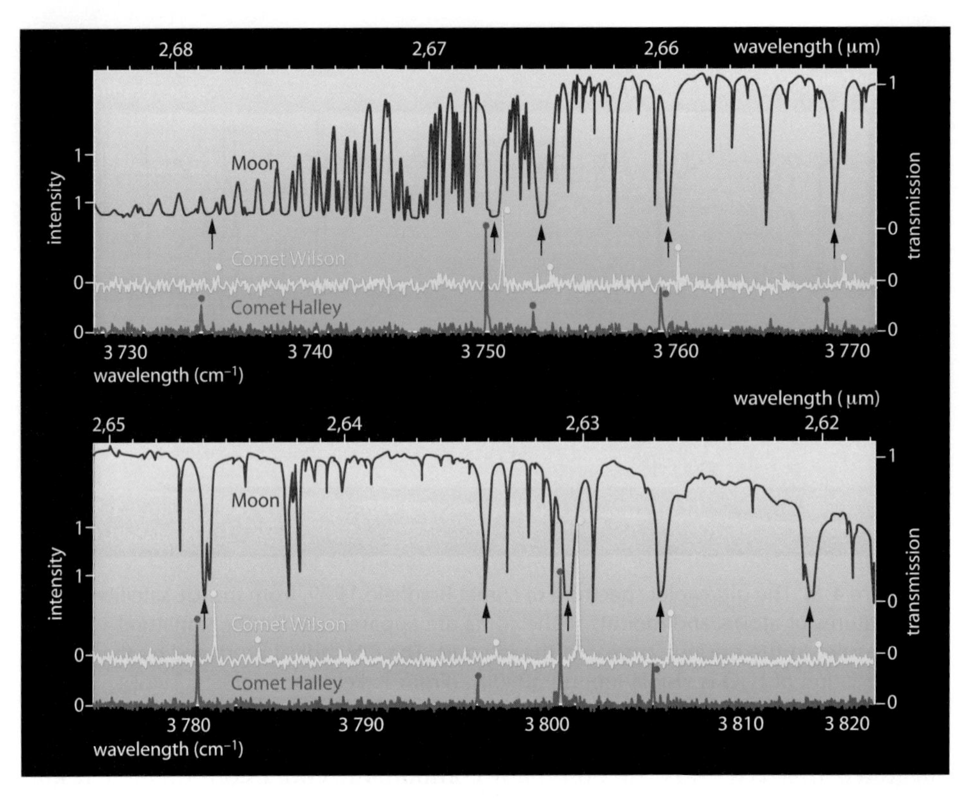

Figure 4.6. The spectrum of H_2O in Halley's comet (26 March 1986) and comet Wilson (12 April 1987), observed from the stratospheric KAO aircraft. The spectrum of the Moon, reproduced above the cometary spectra, shows the transmission of the Earth's atmosphere. The horizontal axis shows the wave number in cm^{-1} – the reciprocal of the wavelength. The lines due to cometary water, marked by dots, are shifted in relation to the absorption lines of atmospheric water (arrowed), as a result of the Doppler effect. The velocity of comet Wilson with reference to that of the Earth was –47 km/s; for Halley, the value was +35 km/s. (From H. Larson *et al.*, 1989.)

together observations by spacecraft and ground-based instruments working at all wavelengths, therefore marked a great step forward in the development of cometary studies.

The most notable discovery was that of water vapour, using infrared spectroscopy, concentrating on the vibrational bands around 2.7 μm (Figure 4.6). The spectral signature of cometary water takes the form of very narrow, well-separated individual vibrational–rotational lines. The emitting mechanism causing these lines is *fluorescence*, whereby infrared solar photons ($\lambda = \lambda_A$) are absorbed by cometary water molecules which are thereby brought to higher energy levels. As they fall back to the fundamental energy levels, they emit a photon at exactly the same absorption wavelength ($\lambda_E = \lambda_A$). Fluorescence favours the infrared part of the spectrum rather than shorter wavelengths

Halley's comet

Halley's comet is undoubtedly the most famous comet in history. Known since ancient times, it was used by astronomer Edmond Halley to determine the true nature of cometary objects. Halley studied its returns of 1456, 1531, 1607 and 1682, and successfully predicted its reappearance of 1758, brilliantly confirming Newton's theory. The comet became famous, and ensured that Halley's name would long be remembered.

The 1910 apparition was a particularly favourable one for observing the comet from Earth, as it passed close by and was seen in the night sky when it was near perihelion. Excellent photographs were taken at the time, as were spectroscopic measurements in visible light. This apparition also caused a great furore among the public, fascinated by the spectacle of the comet's tail, but also fearful of the possible effects of toxic cyanide vapours as the Earth swept through it!

Halley's comet has two characteristics which make it an ideal target for space exploration. On the one hand it is bright, because, with its relatively long period (76 years) it has not yet lost all its reserves of ice during its many perihelion passages. On the other hand its trajectory is accurately known, as a result of the many observations made of it. It was therefore chosen as the target of an unprecedented set of observations from space, to be carried out

Figure 4.7. A photograph of Halley's comet in May 1910. Due to the particularly favourable geometrical configuration at this apparition, images of exceptional quality could be obtained. The planet Venus (below at left) provides some idea of the brightness of the comet. (Photograph taken at Lowell Observatory, Flagstaff, Arizona; courtesy of Paris Observatory at Meudon.)

Figure 4.8. In March 1986 the European Giotto probe approached Halley's comet to within 500 km, and secured the first images of a comet's nucleus.

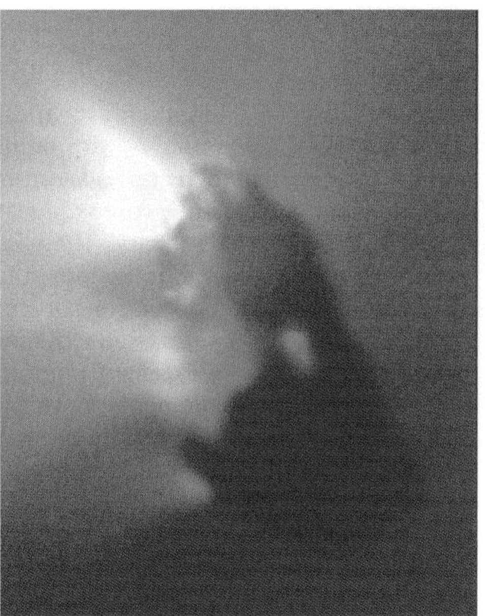

Figure 4.9. The nucleus of Halley's comet as seen through the camera of the Giotto probe as it flew by the comet on 13 March 1986. The two very bright areas are active regions of the nucleus. It is elongated, measuring 15 × 7.5 km × 7.5 km.

during its passage in 1986. Five probes encountered the comet in March 1986, shortly after its perihelion passage: Giotto (European), Vega 1 and Vega 2 (Soviet), and Suisei and Sakigake (Japanese). Giotto approached within 500 km of the comet, securing the first images of a cometary nucleus. This programme was reinforced by an unparalleled effort from ground-based observers. The latest techniques – especially in infrared and millimetre spectroscopy – were brought to bear, and astronomers learned a vast amount in spite of a somewhat unfavourable configuration, the comet being behind the Sun at perihelion on 9 February 1986.

The 1986 passage of Halley's comet marked a great leap forward in our understanding of cometary physics. As well as confirming Whipple's 'dirty snowball' theory, it furnished new and sometimes unexpected information, especially about the nature of the nucleus and the composition of the dust, both of which contained carbonaceous material in great quantity. For the first time, the shape of a cometary nucleus (in this case, an ellipsoid 15 × 7.5 × 7.5 km) was seen, and its albedo of 0.04 precisely measured.

because the solar radiation responsible for the mechanism is stronger there (culminating in the visible). The spectral signatures of water are equally intense in this domain (see Chapter 1).

Credit for the discovery of cometary water goes to Michael Mumma and his team, who used a high-resolution spectrometer (with a resolving power of 10^5) mounted on a telescope carried aloft by NASA's Kuiper Airborne Observatory (KAO). The KAO's operating altitude of 14 km allows astronomers a partial alleviation of the opacity of the atmosphere due to water vapour. To counter the residual terrestrial absorption, the observers chose the time when the comet exhibited its maximum radial motion (along the line of sight from the Earth). At such times, radiation emitted by the celestial body in question is slightly shifted in frequency (or wavelength) due to the Doppler effect (see Chapter 2). Two sets of observations were carried out a few weeks before and after 9 February 1986 – the date of the comet's perihelion passage. The tiny shift in wavelength was sufficient to separate the lines due to cometary water vapour (approximately ten times narrower than the Doppler shift) from those due to terrestrial water vapour (Figure 4.11).

On 6 March, a few months after this discovery, the infrared spectrometer onboard Vega 1 also detected this celebrated water vapour emission line as the probe neared Halley. With a resolving power of 60, the instrument was unable to separate individual emission lines, but had the advantage of being very close to the source of the emissions. The very intense emission band was unambiguously identified.

In both cases, the observed abundances suggested a production rate (based on the number of molecules released into space per second) which agreed perfectly with predictions. Water was therefore definitely the most abundant component of the comet, comprising 80% of its total mass.

Thereafter, other comets were observed from the KAO: for example, comets Wilson (1987), Austin (1989) and Levy (1990). The KAO was no longer available for observing the last two great comets of the twentieth century, Hyakutake (1996) and Hale-Bopp (1997), but fortunately the ISO was operational as Hale–Bopp became visible. In spite of limited opportunities for observing, and the distance of the comet (3 AU or more), due to the exceptional sensitivity of the ISO's instruments, and in particular its SWS spectrometer, near-infrared water vapour spectra of unprecedented quality were transmitted. Similar results were obtained when the still-operational ISO turned its instruments onto comet Hartley 2.

AN ELUSIVE KIND OF ICE

Cometary water ice, rather than water vapour, should have been detected, given that the ice is observable from Earth; but this was not the case. The simple reason is that by the time that a comet becomes bright enough to be observed from the ground, the ice has already sublimated to form the coma, and the nucleus and its

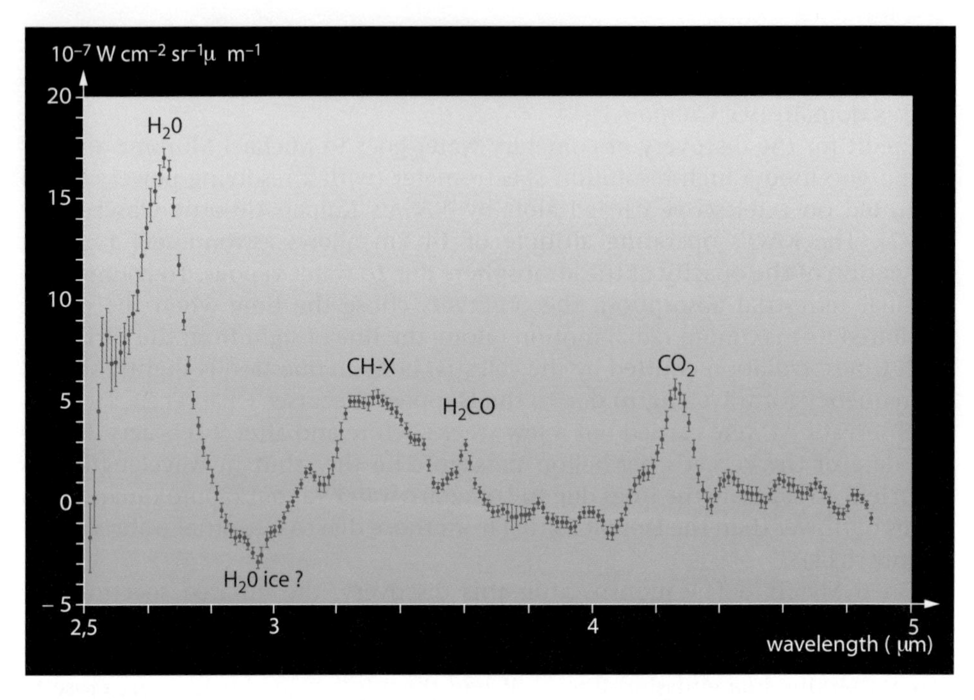

Figure 4.10. The spectrum of Halley's comet observed by the IKS spectrometer onboard the Vega 1 spacecraft. Water ice was detected at wavelengths around 3 μm. (From Combes *et al.*, 1988.)

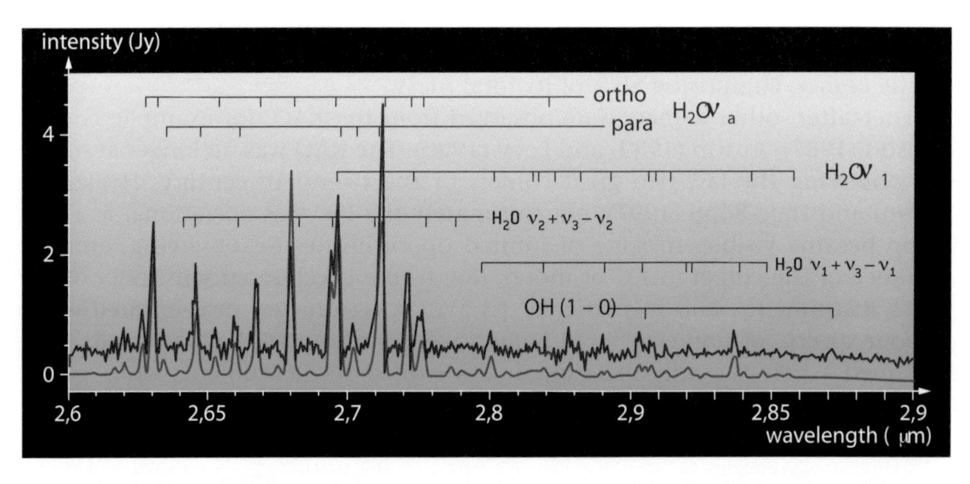

Figure 4.11. The spectrum of comet Hale-Bopp observed by the SWS spectrometer onboard the ISO satellite, in the ν_1 and ν_3 bands of water. The lower curve corresponds to a theoretical spectrum. The relationship between the intensities of the ortho and para water lines corresponds to a formation temperature of 25 K. (From J. Crovisier *et al.*, *Science*, **275**, 1997.)

icy surface are veiled by the dust in the coma. For this reason, astronomers had tried, even before the passage of comet Halley, to detect infrared signatures in more distant comets, which are of course not so easy to observe. Nothing positive came of this for a long time, until, in 1986, the IKS spectrometer onboard Vega 1 managed to detect, with difficulty, water ice on Halley at around 3 μm (Figure 4.10). The ISO made a more convincing detection of cometary water ice, on comet Hale-Bopp, working at 45 μm in the far-infrared.

So, why bother to look for water ice when we know that water vapour is the most abundant of the so-called 'mother molecules' escaping directly from the nucleus? Over and above the confirmation of its existence in the nucleus, the form of its spectral signature can provide us with information on its nature.

As previously stated, water ice may be either amorphous or crystalline, according to the physico-chemical conditions present during its formation. The crystalline form is the signature of relatively high temperatures, while the amorphous form is associated with very low temperatures, such as are often found in the interstellar medium. This is also the case with silicates. Their structure is crystalline in comets, but it tends to be amorphous within the interstellar medium. Cometary water ice seems to be mostly crystalline in nature, like the silicates associated with it. In the case of water, this crystallisation might be expected at temperatures above about 100 K, which are those of the environment of the comets.

WATER ICE ... AND OTHERS

The other components of comets are of two kinds: refractory, or in the form of other ices. How do we detect these other ices in the nucleus? As with water, we observe the products of sublimation: the 'mother molecules'. Before the comet Halley missions, we recognised only the products of dissociation and ionisation, observable in the visible and the ultraviolet, according to the capabilities of the instruments available. When infrared and millimetre spectra became feasible, other mother molecules were within our reach. The first of them was hydrogen cyanide (HCN). Its detection coincided with the first observations by the large millimetre IRAM (Institut de Radioastronomie Millimétrique) antenna in Spain, which, most fortunately, was operational in time for comet Halley's appearance in 1986. Other discoveries followed this, due largely to the passage of comet Hale–Bopp in 1997, and more than twenty cometary molecules are now known. Particular mention should be made of carbon monoxide (CO), identified from its ultraviolet signature in 1976. CO is the most important minor component, with a relative abundance sometimes 20% that of water.

It is very difficult to compile a list of the relative abundances of mother molecules, as the sublimation curves vary considerably from one to another. H_2O and CO represent the two extreme cases, with water the most volatile and CO the least volatile. Comet Hale–Bopp provided an example of this phenomenon. The ratio in the vapour phase of CO/H_2O abundances, measured by the ISO at

Table 4.1. Abundances of mother molecules known to exist in comets.

Molecules	Abundance	Method of observation
H_2O	100	IR, dissociation products (H, OH, O) in UV, visible, radio
CO	2-20[1]	UV, radio
CO_2	3	IR
H_2CO	0.034[1]	Radio, IR
CH_3OH	1-8	Radio, IR
HCOOH	< 0.2	Radio
CH_4	< 1?	IR
NH_3	0.11	Dissociation products (NH, NH_2) in UV, visible
HCN	~ 0.1	Radio
N_2	0.02–0.2	Ionisation product (N_2^+) in the visible
H_2S	~ 0.2	Radio
CS_2	0.1	Dissociation product (CS) in UV
OCS	< 0.3	Radio, IR
SO_2	< 0.001	UV
S_2	0.05[2]	UV

1 These molecules appear to emanate from both the nucleus and a source distributed within the comet's atmosphere (as a result of outgassing of dust and dissociation of more complex molecules). The deduction of their abundances is thus an inexact science.

2 Observed in only one comet. Most of these molecules were also detected in ion or proton form by the mass spectrometers on board Giotto.

The Rosetta mission

The Rosetta mission was one of those selected by the European Space Agency in the early 1980s as part of its long-term Horizon 2000 programme. Conceived initially with a view to returning a comet sample, the mission was redefined in the 1990s as an observational one, the aim being to orbit a comet and study its various phases of activity as it approached perihelion. Moreover, a lander would touch down on the nucleus and measure its physico-chemical characteristics *in situ*. The cometary target would have to be relatively inactive, in order not to damage the instruments. The first comet to be selected was periodic comet Wirtanen. However, the January 2003 launch window was lost due to an accident with an Ariane 5 rocket two months earlier, and another observational mission was planned, this time with periodic comet Churyumov–Gerasimenko. The mission finally got under way in February 2004, and will last twelve years, with surveillance of the comet throughout its burgeoning activity until the time of perihelion passage in 2015.

The Rosetta probe consists of an orbiter which will constantly monitor the nucleus with a battery of instruments, and a lander carrying a vast array of experiments. On the orbiter there will be a camera, spectrometers operating in ultraviolet, visible, infrared and submillimetre wavelengths, mass spectrometers for the study of gas and dust particles, dust analysers, and instruments to investigate the interaction of the comet and the solar wind. With all these, the characteristics of the surface will be revealed and its activity scrutinised. The products of outgassing will be identified, as will the composition of the inner coma. Among the many instruments on board the lander will be cameras, spectrometers to measure the composition of the surface material, and a magnetometer. A radar device fixed to the orbiter will attempt to visualise the internal structure of the nucleus.

The long journey of the Rosetta probe, with its arsenal of experiments, will encompass more than a decade observing its target, adding to our knowledge of the physical and chemical processes governing cometary activity. New insights into the genesis of comets will also broaden our understanding of the birth of the solar system.

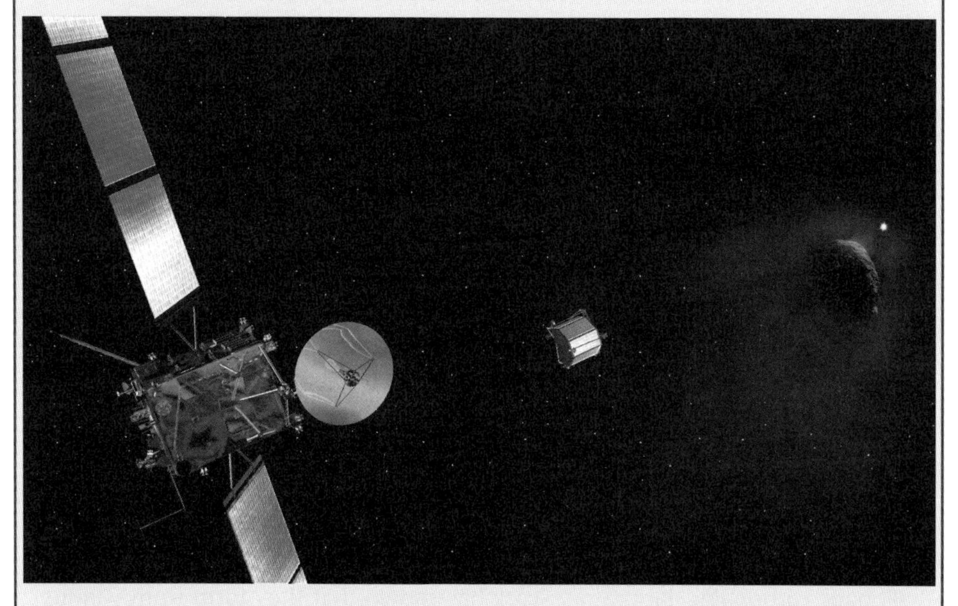

Figure 4.12. The Rosetta probe will be put into orbit around a comet, and is programmed to study the various phases of its activity as it approaches perihelion. A descent module will land upon the nucleus to make *in situ* measurements of its physico-chemical properties.

distances of 4.7 and 3 AU respectively, was much greater at a great heliocentric distance, with both values higher than they would be at 1 AU. In order to estimate quantitatively the relative abundances of volatile molecules such as CO by observing the rate of production at distances of the order of 1 AU, we have to be able to estimate how a molecule like CO is capable of diffusing through water ice. This presupposes a very accurate knowledge of the fine structure of a cometary nucleus, and we are far from achieving this. Another question also arises. Are other kinds of molecule present in the form of pure ice or as clathrates? Clathrates, we will recall, are matrices of crystalline water ice able to entrap atoms and other molecules, in quantities of one external element for several water molecules. New light will undoubtedly be thrown on this question in 2014, when the Rosetta mission explores comet Churyumov–Gerasimenko (see below, and fact box, p. 88).

Refractory materials: silicates and organic matter

The refractory phase presents two components – silicate and carbonaceous material – the nature of which has been revealed to us due to studies of Halley's comet. Earth-based observations at around 10 μm had already shown that dust in comets is composed largely of silicates. More accurate measurements of Halley by the KAO and the IKS-Vega spectrometer indicated that the substance in question was olivine, a crystalline mineral. Ten years later, the ISO brought a further insight: the far-infrared spectrum of comet Hale-Bopp shows all the signatures of forsterite (Mg_2SiO_4) – a particular type of olivine rich in magnesium.

As for carbonaceous material: their discovery in 1986 on the surface of comet Halley's nucleus by spaceprobes led to a veritable revolution. Three types of observation led to their detection. First, the camera on board Giotto showed that the surface was dark – very dark. Its albedo was 0·04, equalling the darkest surfaces known on Earth. Then, the IKS-Vega spectrometer found a wide spectral signature at 3.3–3.4 μm – a signature already known from several observations of the interstellar medium. According to the astronomers, it corresponded to complex organic constituents, both aliphatic (based on a carbon chain) and aromatic (complex cyclic molecules organised around carbon benzene rings). The latter have been detected in quantity in space, in the form of PAHs (polycyclic aromatic hydrocarbons), and they contain a considerable amount (of the order of 30%) of the total carbon. As for the aliphatic carbonates, laboratory work has shown that they form from various mixtures of ices (H_2O, CH_4, NH_3, and so on) when subjected to strong irradiation (UV, or very energetic cosmic rays). Another experiment confirmed this result. Mass spectrometers on Giotto and Vega detected the presence of large amounts of complex organic molecules with atomic masses of several tens of units. They also showed that the 'light' elements were there in abundance, thus confirming the primordial nature of cometary material, which captured light elements (with the exception of hydrogen and helium) within its ices.

Why did the discovery of carbonaceous material create such a stir at the time? Because, for some, the presence of organic matter on comets opened up new vistas in exobiology. Humans have long wondered about the origin of life on this planet. We do not know how it came to be, but we know that its appearance was linked to water. Since water has long been thought to be an ingredient of comets, certain researchers have suggested that life – indeed, living organisms – might have perhaps been brought to Earth on comets – the 'panspermia' hypothesis. In 1986, some scientists, and in particular the astronomer Fred Hoyle, saw the cometary 3-μm spectral signature as proof of the presence of molecules of biological origin. It soon transpired, however, that the idea was unconvincing, as many abiotic organic molecules also produce such a signature. However, the discovery of organic material on Halley's comet remains interesting, because even if the molecules turn out to be abiotic, they could still be 'prebiotic'; that is, capable of further chemical transmutations – transmutations which may have spelled the beginnings of life on Earth.

As well as the problem of 'selective sublimation', there is the question of some mother molecules such as CO and H_2CO having an alternative origin. They are thought to derive partially from the sublimation of carbonaceous grains ejected with the dust and gas. Estimation of element abundances within cometary nuclei is not an exact science!

COMETARY MATTER AND INTERSTELLAR MATTER

Did comets form from interstellar matter? There are several indications to confirm this hypothesis. First, let us compare cometary ice with interstellar ice. The analogy is striking. Not only are they similar in nature, but also in their relative abundance, bearing in mind the reservations expressed elsewhere on the subject of the relative abundances of cometary ices. If we extend the comparison to refractory elements (carbonaceous material, silicates and metals), the same remarkable agreement is found.

Comparing the types of gases detected in comets with those observed in the interstellar medium, we arrive at the same conclusion. It is a fact that all the mother molecules present in comets have also been observed in the interstellar medium. The list of interstellar gases is long, and includes a great number of radicals and unstable molecular ions, resulting from the ultraviolet irradiation of molecular clouds by stars. Within these dense and very cold clouds are neutral molecules, and it is this type of cloud which may contract, as a result of its own mass, into new stars - a scenario analogous to our theories about the formation of the solar system. The analogy also holds for more complex molecules such as polycyclic aromatic hydrocarbons (PAHs), on the borderline between heavy molecules and interstellar grains. PAHs may be observed because they re-emit, in the infrared, UV photons absorbed from nearby sources.

It is natural to see, within this global analogy, confirmation of a general model of an 'interstellar cycle', whereby dust finds its way from the interstellar medium into the material surrounding infant stars, and is then blown back into the interstellar medium during the last stages of those stars' lives, to be recycled into a succeeding generation of new stars.

WATER: HISTORIAN OF THE COMETS

In our research into the origin of comets, the study of their water has supplied us with two essential tools – the first derived from the study of heavy water (HDO), and the second from measurements of the ratio of the ortho and para states of H_2O.

We have already seen in Chapter 1 that the study of heavy water and its relative abundance compared with H_2O can provide information about the temperature of the formation of the milieu in which they are found. At low temperatures, ion–molecule reactions ($H_3^+ + HD \rightarrow H_2D^+ + H_2$) and molecule–molecule reactions ($HD + H_2O \rightarrow HDO + H_2$) favour the production of H_2D^+, precursor of HDO, and HDO itself. These reactions imply enrichment (in D) of ices, and this is observed in the interstellar medium. The deuterium excess in the ice content of a body is therefore an indicator of the temperature at which the body formed. Note that this phenomenon is also valid for other hydrogen-based ices: HCN, CH_4, and so on. (We shall return to this point in more detail in Chapter 6.)

Three measurements of the HDO:H_2O ratio in comets have been obtained – the first with the mass spectrometer of the Giotto probe as it swept past comet

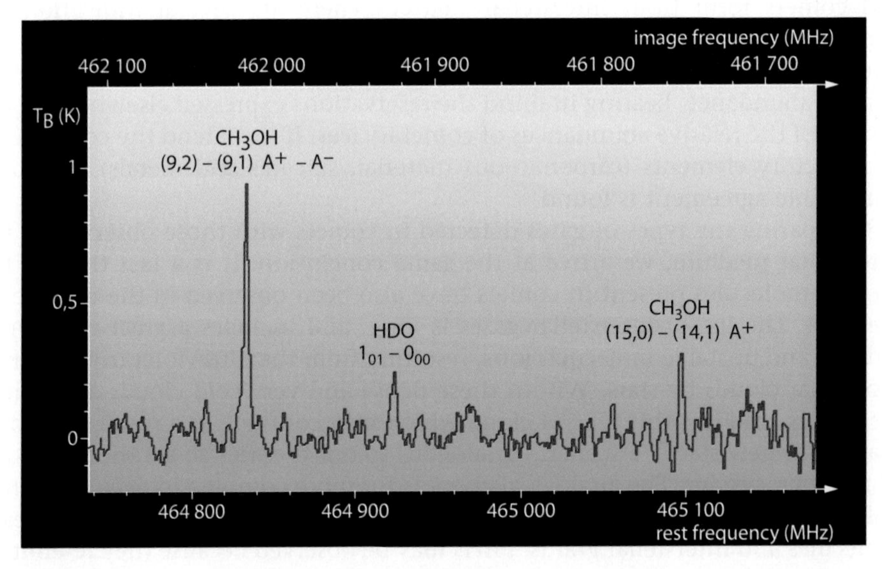

Figure 4.13. The spectrum of comet Hyakutake as observed from the Caltech Submillimeter Observatory (CSO). This observation revealed the D:H ratio within the comet, by comparison with the measurement of the abundance of H_2O. (From D. Bockelée-Morvan *et al.*, 1998.)

Comet Hale–Bopp

Discovered simultaneously by two amateur astronomers, Alan Hale and Thomas Bopp, on 28 July 1993, comet Hale–Bopp was, at the time, still 7.1 AU from the Sun. With a nucleus more than 50 km across it became one of the most brilliant comets in history. Due to a favourable geometrical configuration it was visible to the unaided eye for nearly two months as it rounded the Sun, perihelion passage occurring on 1 April 1997. At perihelion it was 0.9 AU from the Sun and was ejecting approximately 10^{13} water molecules per second – ten times the production of Halley's comet. Comet Hale–Bopp was intensively observed, both from the ground and by the Hubble Space Telescope and the ISO, which was still in operation at that time.

Observations indicated multiple jet structures within the coma, and the first example in a comet of a rectilinear tail of atomic sodium. Due to its exceptional brightness, its gas and dust emissions could be studied at considerable distances from the Sun, both before and after perihelion. Various new molecules were discovered by infrared, millimetre and submillimetre spectro-

Figure 4.14. Comet Hale–Bopp was one of the most brilliant comets of the twentieth century. It was the target of an intensive programme of observations, from the ground, from the Hubble Space Telescope and by the ISO satellite. Its passage led to definitive advances in cometary science. (Photograph courtesy M. Jourdain de Muizon.)

scopy; in particular, astronomers measured the D:H ratio not only in water but also in HCN, having detected molecules of H_2O, HDO, HCN and DCN. Among ISO observations were a detailed study of infrared water bands, the identification of water ice and of silicates constituting cometary dust – in this case, forsterite, a variety of olivine rich in magnesium and found in circumstellar envelopes around young and evolved stars.

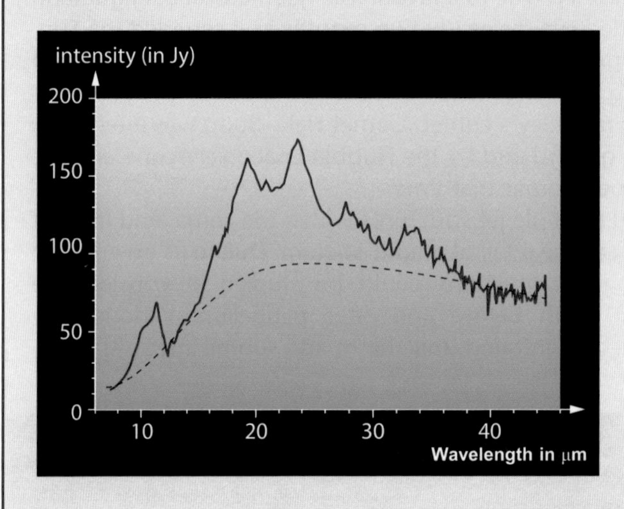

Figure 4.15. The ISO–SWS spectrum of comet Hale–Bopp between 7 and 45 μm shows emissions due to dust, and reveals the signature of forsterite (Mg_2SiO_4) – a crystalline olivine rich in magnesium. The dashed curve shows the emission of the nucleus minus the gas and dust. (From J. Crovisier *et al.*, *Science*, **275**, 1997.)

Halley, and the other two, based on measurements of comets Hyakutake and Hale–Bopp, of an HDO transition in the submillimetre (Figure 4.13). In all three cases, the D:H ratio was found to be $3{:}10^{-4}$ – twice that of the Earth's oceans and approximately fifteen times greater than in the protosolar nebula. This result reinforces our global picture of the formation of the solar system: comets, being small bodies formed from planetesimals, could not have accreted any protosolar gas. We also learn that the water in the Earth's oceans is not entirely of cometary origin, as some have proposed. As the D:H ratio of the oceans is half that of comets, Earth's atmosphere must have had a different origin, at least in part. As will be seen in Chapter 7, the atmosphere came from both within (especially through volcanism) and without (as a result of infalling asteroids, which are less rich in deuterium than are comets).

Lastly, it is worth mentioning measurements of the ortho:para ratio of H_2O in the make-up of comets. In Chapter 1, we saw that the two types have spectral signatures at slightly different frequencies (and therefore wavelengths), so that independent analysis of the two different kinds can be carried out, and the relative abundance revealed. This relative abundance, unchanged since their origin, depends on the temperature obtaining at the time the molecules were formed. The experiment has been performed in the case of three comets, using infrared spectroscopy at 2.7 μm. It is not an easy measurement, requiring high spectral resolution to separate individual emission lines, and high sensitivity in order to measure relative abundances. For comet Halley, KAO measurements had been used; but ten years later, the ISO was capable of much greater accuracy of the ortho:para ratio in comets Hale–Bopp and Hartley 2. Observed temperatures, of the order of 25–35 K, confirmed that comets originated at low temperatures, in the interstellar medium or the outermost reaches of the solar system.

The origin of comets

Edmond Halley's work showed that comets are not of atmospheric origin. Their very eccentric orbits can take them out beyond the paths of the known planets. Where do they come from? In 1950, Dutch astronomer Jan Oort, taking into account planetary perturbations, determined the initial orbits of about twenty comets before they had interacted with planets. He demonstrated the existence of a huge cloud of comets, in a halo between 50,000 and 100,000 AU from the Sun. Since its discovery it has been called the Oort Cloud. Later work by astronomer Brian Marsden confirmed the hypothesis. According to modern estimates, all non-periodic or long-period new comets originate in the Oort Cloud 'reservoir', and often move in orbits highly inclined to the plane of the ecliptic. The cloud may contain some 10^{11} comets, but only a tiny fraction of this number will be drawn inwards into the solar system, as the gravitational effects of nearby stars, or even interstellar forces, come into play. Then, new and unexpected comets will make their occasional appearances. These are often bright objects, because they are approaching the Sun for the first time and their mantles of ice have yet to be stripped from them.

How was the Oort Cloud created? Local densities are such that no comet could have formed there. The application of celestial mechanics suggests that the Oort comets originally formed at a few tens of AU from the Sun, near the orbits of Uranus and Neptune, and that perturbations from Jupiter caused the comets to be ejected into the Oort Cloud.

As well as the Oort comets, there are also short-period comets, with orbits not much inclined to the plane of the ecliptic. They are often less active than new comets, since, with their orbits of only a few years, they have largely lost their coating of ice during successive passages near the Sun. Where do these bodies originate? Here again, celestial mechanics can come to our aid, and we can retrace the history of these small solar system bodies since their origin. The consensus among comet experts nowadays is that comets were born in a region not far from the ecliptic, about 30–100 AU from the Sun. This region is known as the Edgeworth–Kuiper Belt, after the two astronomers who put forward the hypothesis (although it is usually referred to as the Kuiper Belt). We shall later return to this new frontier of the solar system, where, in the last decade, astronomers have discovered a multitude of small, asteroid-sized bodies moving beyond the orbit of Neptune. Well over 1,000 of these trans-Neptunian objects are now identified, and the list continues to grow.

Most of our information about comets is based on members of the Oort Cloud, simply because they are brighter and therefore more easily observed. One of the challenges facing cometary physicists in years to come will be to obtain better information about Kuiper Belt comets. A decisive contribution to this work will be made early in the next decade by the European Rosetta mission, visiting comet Churyumov–Gerasimenko.

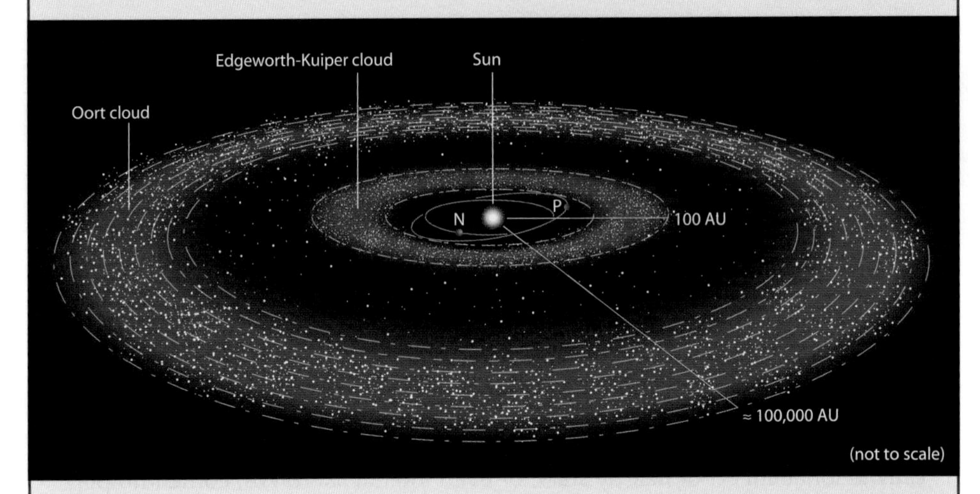

Figure 4.16. Where do comets come from? The two reservoirs of comets in the solar system are the Oort Cloud, between 10^4 and 10^5 AU from the Sun, and the Edgeworth–Kuiper Belt, 30–100 AU from the Sun. The diagram shows a section through these belts in the ecliptic plane. While the Edgeworth–Kuiper Belt is confined to the vicinity of this plane, the Oort Cloud takes the form of a shell extending to high heliocentric latitudes.

SPACE EXPLORATION OF COMETS: RECENT RESULTS AND FUTURE PROJECTS

In the wake of the ISO mission, several comets have been the targets of spacecraft such as Spitzer, SWAS and Odin. In particular, the systematic observation of the strong submillimetre water vapour transition at 557 GHz has increased our knowledge of the activity of many comets, as have studies of thermodynamic conditions within their inner comae.

In July 2005, NASA carried out an ambitious mission when the Deep Impact spacecraft sent a projectile, weighing several hundred kilogrammes and made of pure copper, to collide with comet Tempel 1. Scientists hope that observations of the ensuing crater and the ejected material will tell us more about the nature of the interiors of cometary nuclei. Another very promising American mission is Stardust. Launched in 1999, Stardust flew by the nucleus of comet Wild 2 in 2004, gathering samples of both cometary and interstellar dust. The samples arrived safely on Earth in January 2006, and their analysis over the next few years will provide a great deal of information on the refractory component of cometary matter.

Finally, the ambitious Rosetta mission, created by the European Space Agency, will act as a veritable space laboratory, analysing a comet's nucleus as it approaches the Sun during the next decade (see fact box, p. 88). Rosetta was launched in February 2004, and will encounter comet Churyumov–Gerasimenko when it is still at a distance of several AU from the Sun. It will place a module on the comet's surface for *in situ* analysis of its constituents, and will accompany it on its journey as it nears the Sun, recording the development of its activity and the growth of its coma.

5

Water in the outer solar system

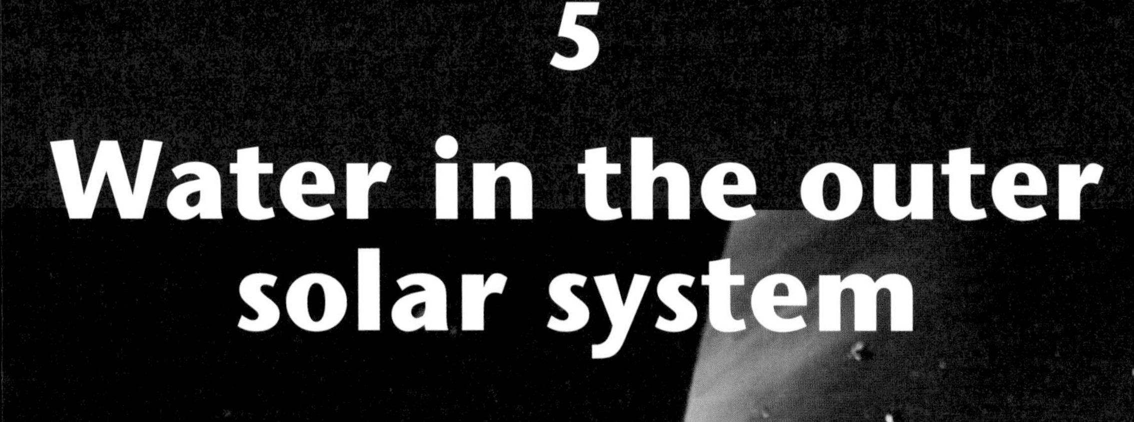

The giant planets, their satellites and rings, and the trans-Neptunian objects seem to be composed of icy nuclei, with water ice as the principal ingredient, in the manner of comets. But what is the origin of the water vapour discovered in the atmospheres of the giants? Did it come from the rings, or the icy satellites, or from a flux of micrometeoroids continuously drawn in by the planets' gravitational fields? How do we explain the similarity of the atmospheres of Pluto and Triton? Have they a common origin?

There is a wide variety of objects in the outer solar system, ranging from its biggest planets – the gas giants and ice giants – to the flux of micrometeoroids sweeping through interplanetary space. At 3 AU from the Sun, we cross the ice line (see Chapter 3). The satellites, rings, trans-Neptunian objects, and the planetesimals which made up the nuclei of the giant planets, were therefore made of ice – mostly water ice – like comets.

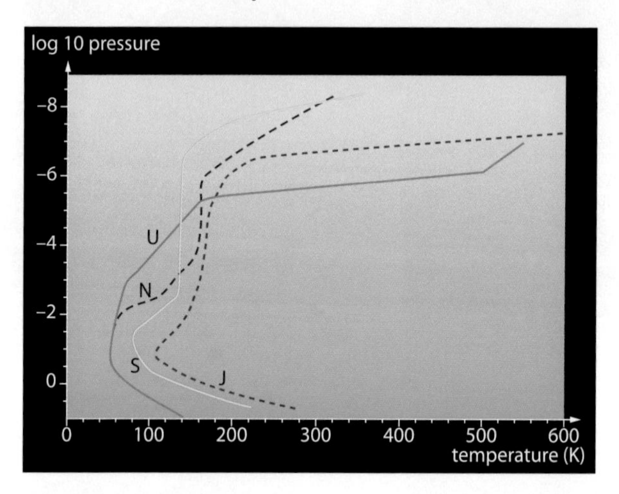

Figure 5.1. Vertical thermal profiles of the giant planets. The profile for Jupiter is based on measurements taken by the Galileo descent probe. The profiles of the other three planets are deduced from Voyager measurements. The low temperature at the tropopause causes the condensation of several molecules, one of which is water. (From T. Encrenaz, *Astron. Astrophys. Rev.*, **9**, 1999.)

The giant planets and their retinue of regular satellites are rather like miniature solar systems: the accretion of the surrounding sub-nebulae led to the formation of a disc within which both rings and regular satellites accreted. Later, the other satellites, which have high inclinations and very elliptical orbits, joined the systems, captured by the planets' gravitational fields.

As will be seen, in order to remain stable and maintain internal cohesion a satellite must be beyond the Roche limit, at a distance of 2.5 times the planetary radius. Within this limit, rings predominate, as differing tidal effects caused by the gravitational field tend to separate out the different parts of the satellite.

We owe much of our current knowledge of the giant planets and their systems to a series of American spaceprobes. The Pioneer 10 Jupiter mission (1972), and Pioneer 11 to Jupiter and Saturn (1973) came first; then, Voyagers 1 and 2, whose

---

An artist's impression of Saturn as seen from its rings.

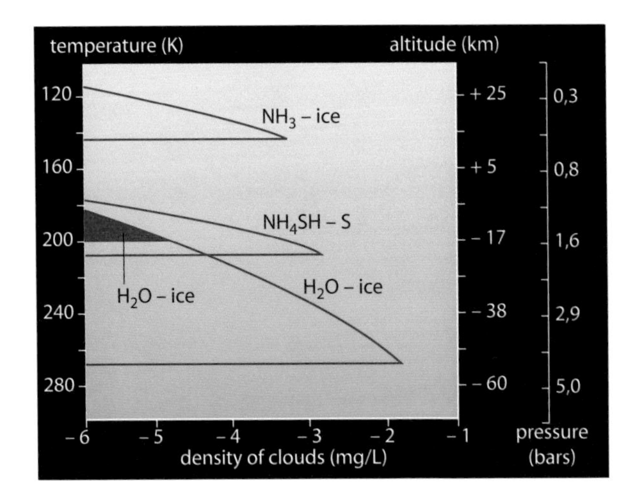

Figure 5.2. Cloud structure on Jupiter. (From R. Prinn and T. Owen, *Jupiter*, ed. T. Gehrels, University of Arizona Press, 1976.)

batteries of assorted instruments sent us such a mine of information, still being studied to this day. The Voyagers flew by Jupiter in 1979, and encountered Saturn in 1980 (Voyager 1) and 1981 (Voyager 2). Voyager 2 reached Uranus in 1986 and Neptune in 1989. A decade later, Jupiter was the target for the Galileo mission, consisting of an orbiter and a descent module. Galileo arrived in 1995, and the orbiter ended its mission in 2003. Finally, the ESA/NASA Cassini spaceprobe, launched in 1995, was designed to carry out exploration of the saturnian system, which its American orbiter will investigate for eight years. On 14 January 2005, the descent module, Huygens, developed by ESA, successfully landed on the surface of Titan.

THE ATMOSPHERES OF THE GIANT PLANETS

As we saw in Chapter 3, the atmospheres of the giants are dominated by hydrogen. Minor constituents are therefore all hydrogenated (CH_4, NH_3, and so on). These materials have been detected in the deep atmospheres of Jupiter and Saturn at pressures greater than a few tens of bars. As well as methane and ammonia, the water molecule is present, as are less abundant types such as PH_3, GeH_4 and AsH_3. However, with the exception of methane, these types have not been confirmed on Uranus and Neptune, because of their lower atmospheric temperatures. The fact is that most types will condense at levels too deep to be observable. Cloud layers identified on Jupiter and Saturn are NH_3, at 0.5 bar; NH_4SH, at around 1–2 bars; and H_2O at a pressure of a few bars. In the atmospheres of Uranus and Neptune, H_2S is suspected at a pressure of 3 bars, and NH_3 and H_2O at about 10 bars. The atmospheres of the giants are characterised by a troposphere, within which temperature falls with increasing altitude, until a minimum is reached at the tropopause (as on Earth). Here, pressure is around 100 millibars, and the temperature is, on Jupiter, 110 K, on Saturn, 90 K, and on

Figure 5.3. Jupiter from Voyager 1. Water is present in Jupiter's lower troposphere. It has been observed by means of infrared spectroscopy in the 5-μm window. The first observations were carried out from the ground and by the KAO, and were later confirmed by the Voyager probes and Galileo. However, it was also revealed that although water is present it is there in much smaller quantities than predicted by the accretion theory – no doubt for some meteorological reasons.

Uranus and Neptune, 50 K. Above this level, the temperature rises again in the stratosphere, which contains other minor constituents – among them several hydrocarbons, products of the photodissociation of methane (Figure 5.1). Among other less obvious ingredients here is water vapour.

WATER AND THE GIANT PLANETS

On the giant planets, in the presence of so much hydrogen, oxygen ought to be found as part of H_2O. The O:H ratio could then be deduced directly from the value H_2O/H_2, if we allow that the oxygen is mostly in the form of H_2O and the hydrogen essentially H_2. According to the accretion theory, the O:H ratio should be three times greater than the cosmic value (see Chapter 2). But what do we find

Figure 5.4. Measurements of the quantity of water vapour in the atmosphere of Saturn show that there is far less than is predicted by the accretion theory. As with Jupiter, this phenomenon can be explained by an atmospheric circulation model involving intense convection, which would hinder the measurement of the O:H ratio in the tropospheres of the two planets. The O/H value representative of the interiors of the planets is reached only at depths that at present are unobservable.

in reality? Water is certainly present in the lower troposphere of Jupiter. It was observed using infrared spectroscopy in the 5-μm spectral window, which allows observation at depths of the order of several bars. The first observations were carried out both from the ground and on the KAO, and were confirmed by the Voyager and Galileo probes. However, a surprise awaited the observers. Water was present, but in much smaller quantities than had been predicted.

It is not easy to measure the abundance of water in the lower atmospheric layers of giant planets. This water – in the form of vapour, and at temperatures above 250–300 K – ought to condense at a higher altitude in the upper troposphere. However, it can react with other atmospheric ingredients to form compounds: for example, H_2O can react with NH_3 to form a cloud of NH_4OH (ammonium hydroxide), known to condense at about 210 K. A photochemical model allows us to estimate the quantity of water vapour present at each atmospheric level. It transpires from these measurements that the abundance of water vapour is far less (by at least two orders of magnitude) than predicted by the model. ISO observations of Saturn taken at the same wavelength produced a similar result.

How can this phenomenon be explained? Planetologists link it to atmospheric circulation. Jupiter and, to a lesser extent, Saturn, exhibit an intense convective circulation, responsible for the belts and zones visible in their atmospheres. It is generally thought that the zones are areas where currents ascend, while the belts are areas of subsidence. On a smaller scale, there are 'hot spots' in the belts where

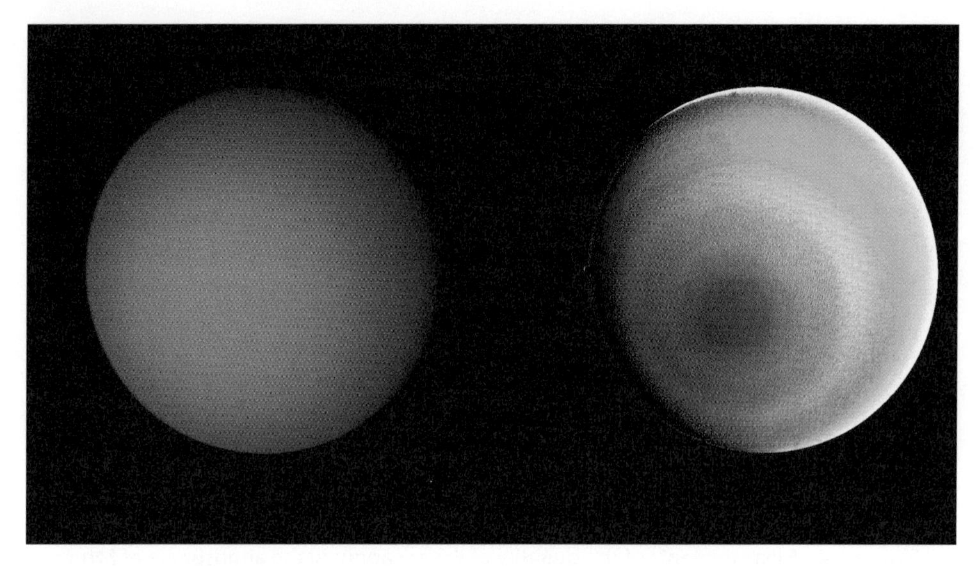

Figure 5.5. Uranus from Voyager 2. The planet appears completely featureless, although the false-colour image increases the contrast. The H_2O molecule is as yet undetected in the tropospheres of Uranus and Neptune.

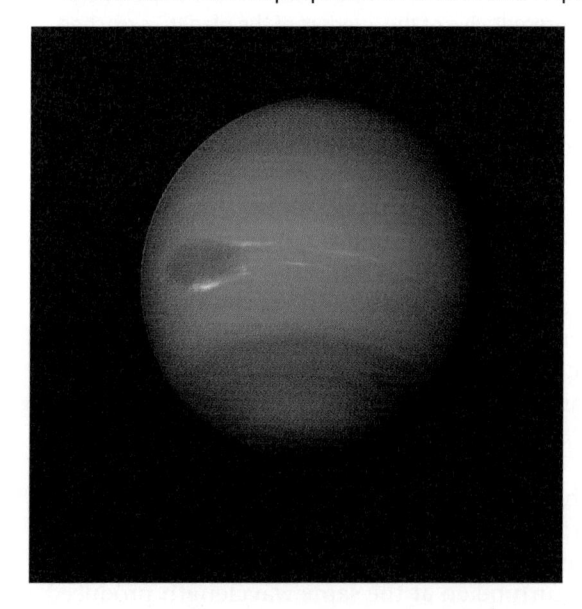

Figure 5.6. Neptune from Voyager 2. The intense colour of Neptune has captured the imagination of observers, and it rivals the Earth as the 'blue planet'. The image shows, for the first time, the existence of a large dark spot and other smaller spots, each proceeding around the planet at a slightly different speed from the others. Ten years later, images obtained by the Hubble Space Telescope showed that the great dark spot had disappeared. This dynamic structure, reminiscent of Jupiter's Red Spot, was not as stable as its famous counterpart.

infrared flux from underlying warmer layers is particularly marked. Galileo's descent probe penetrated one of these regions. The interplay of convective currents causes the ascending zones to be enriched with condensable components, while subsident regions become drier, explaining the low amounts of NH_3 and H_2O measured by the probe. These measurements were confirmed by the infrared spectrometer on the Galileo orbiter, and by the ISO. It is not possible with our present capabilities, therefore, to measure the O:H ratio deep within the jovian or saturnian atmospheres. However, measurement of this parameter is crucial if we are to test the different models of the formation of the giant planets. To be able to measure the O:H ratio deep within Jupiter's atmosphere, we need to plumb its depths, where pressures exceed 20 bars (the value reached by Galileo). This will be the objective of the American Juno mission, now in preparation at NASA, and due to be launched in 2010 or thereabouts. Juno will enter a polar orbit around Jupiter, and among its instruments will be a radio probe capable of receiving signals from layers deep within the planet.

In the case of Uranus and Neptune, the problem of measuring the O:H ratio is even greater. Because they are so cold, H_2O condenses on these two planets at even deeper levels, corresponding to a pressure of about 10 bars. These levels cannot be directly observed, so that the H_2O molecule is still undetected within the tropospheres of Uranus and Neptune. It remains to be said that indirect measurements, based upon the planets' millimetre radiation, suggest a possible H_2O contribution to absorption in the spectrum. However, direct confirmation is needed.

An external source

It may have been no surprise to astronomers that H_2O absorption lines were present in the infrared spectra of Jupiter and Saturn; but they were very surprised when the ISO detected water vapour emission lines in the spectrum of Uranus. With a minimum temperature of 50 K at the tropopause, the atmosphere of Uranus is globally very cold – so cold, moreover, that nobody had imagined that water, in the form of water vapour, might be found there. In the wake of this unexpected discovery, water vapour was sought – and found – on the other three giant planets and on Titan (Figure 5.8). Without delay, another oxygen-bearing molecule, CO_2, already known on Titan, was sought on these planets. It was expected to be there, as a result of photochemical reactions in the presence of H_2O and CO, carbon monoxide having already been detected on Jupiter, Saturn and Neptune before being found on Uranus.

Observation of H_2O emission lines suggested that they originated in the stratosphere, where temperature rises with altitude. The conclusion must be that the water has not risen from the interior, since it would have had to condense out at the level of the 'cold trap' at the tropopause. So, it must have come from outside. But from where? Two possible origins are, firstly, a local source, the giant planets' rings and icy satellites; secondly, some interplanetary source, a flux of micrometeoroids continuously drawn in by the planets' gravitational fields. How do we know which of these two possible sources is the more probable? If the first hypothesis is correct, then Saturn's H_2O flux would be greater because of its rings. However, this value is of the same order of magnitude for all four giants – perhaps a little weaker in the case of Uranus – and for Titan. However, Titan's

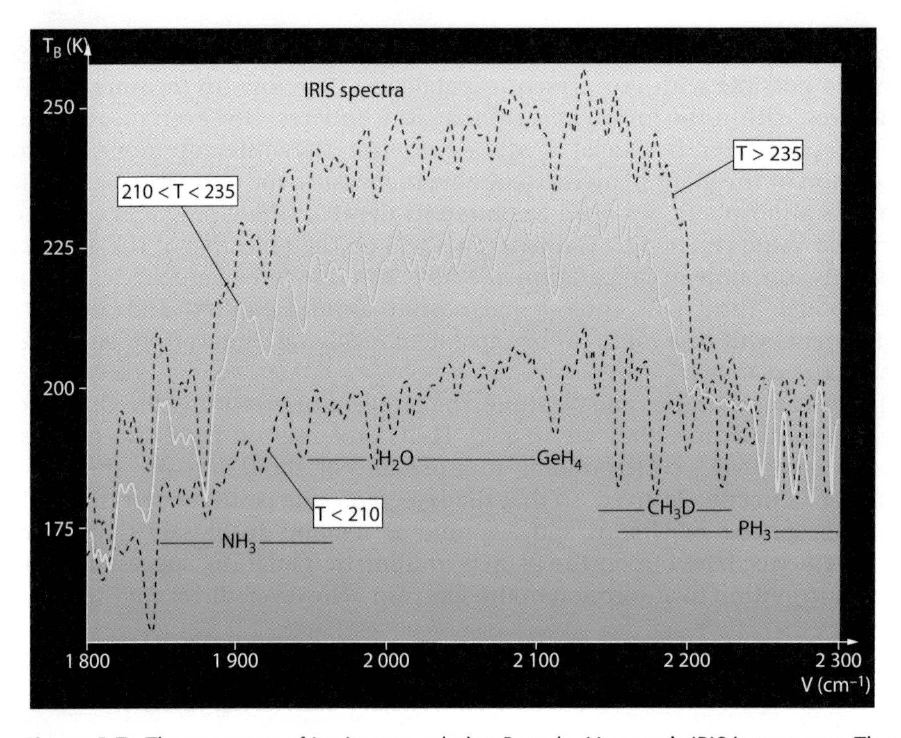

Figure 5.7. The spectrum of Jupiter recorded at 5 μm by Voyager's IRIS instrument. The region between 1800 and 2300 cm^{-1} (λ = 4.5–5.2 μm) represents, for the giant planets as for the Earth's atmosphere, a window for observing radiation from deeper layers – a result of the absence of strong absorption by atmospheric gases (especially methane). This is therefore a favoured spectral domain for research into minor atmospheric constituents in the deep tropospheres of the giant planets (NH_3, PH_3, GeH_4, H_2O). (From P. Drossart *et al., Icarus, 49,* 1982.)

gravitational field is much weaker than that of the giant worlds, and it attracts far less strongly.

Recent studies have shown that, in the case of Jupiter, stratospheric water and carbon dioxide are undoubtedly products of its collision with comet Shoemaker–Levy 9 in 1994. We know that this remarkable event caused the formation of new molecules, such as H_2O and CO, in the stratosphere. Certain other molecules, including CO, CS and OCS, have been found in the millimetre region, over several years; and the water observed in the infrared by the ISO, and by the SWAS and Odin satellites in the submillimetre, may share the same origin.

As far as Saturn and Titan are concerned, the rings and the icy satellites might well be the main sources of H_2O. As for Uranus and Neptune, which lack some large supply in their vicinity, they may have drawn in with their gravity some of the flux of micrometeoroids continuously present throughout the outer solar system. This flux was revealed by the Pioneer 10 and Pioneer 11 probes. The micrometeoroids' chemical composition should be like that of comets, some of

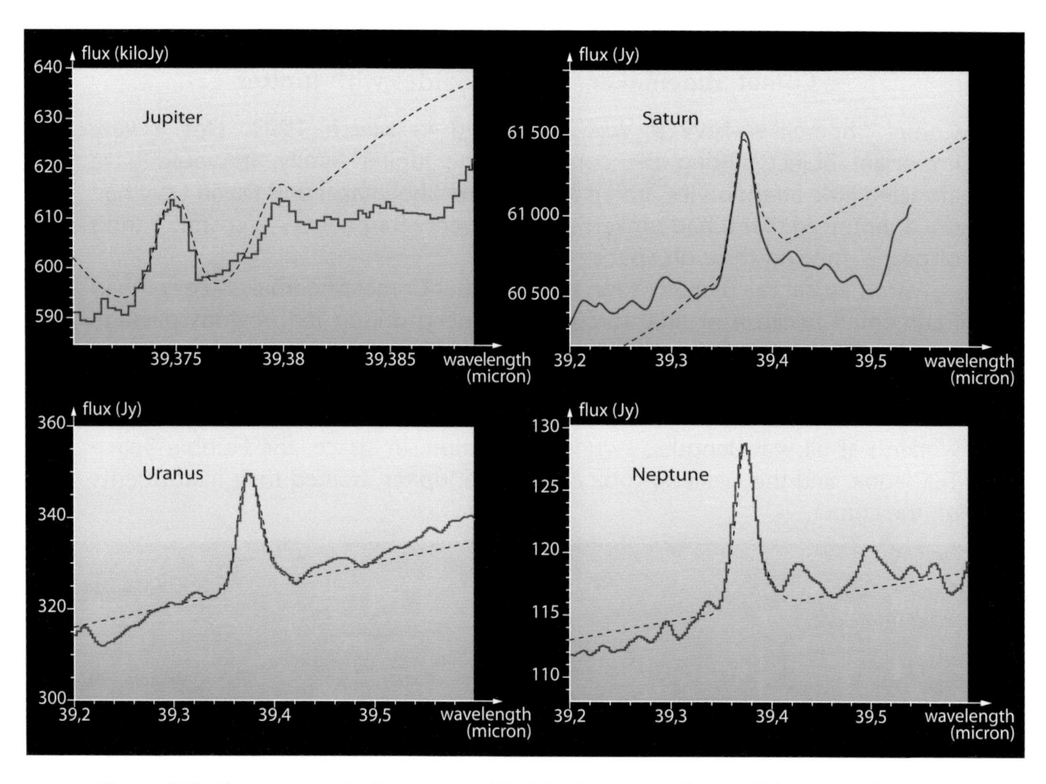

Figure 5.8. Spectroscopic signatures of H_2O in the stratospheres of the giant planets, as observed by the ISO. The observed lines (in red) appear as emission lines, because they are due to the stratosphere, where temperature increases with altitude. The dashed lines represent the spectra predicted by theoretical models. (From H. Feuchgruber *et al.*, *Nature, 389*, 159, 1997.)

which formed near the orbits of Uranus and Neptune. H_2O should be the main component, with some CO and possibly CO_2. The carbon dioxide observed in the planets' stratospheres could come either directly from outside, like H_2O, or be formed locally as a result of photochemical reactions involving H_2O and CO and perhaps CH_3 radicals as products of the photodissociation of methane.

If these hypotheses are correct, then the constancy of the value for the H_2O flux received by the four planets and Titan must be fortuitous. The question remains open. Results from Cassini at Saturn, and of the gas giants by Herschel in the submillimetre domain, will provide decisive new clues to help answer that question.

SATELLITES OF THE OUTER SOLAR SYSTEM

The satellites of the outer solar system exhibit a great variety of surfaces, compositions and physical conditions. Of the Galileans, for example, Io – coated

Comet Shoemaker–Levy 9 collides with Jupiter

Comet Shoemaker–Levy 9 was discovered in March 1993. Not a very important object among the comets of the Jupiter family, it would have attracted little attention if Carolyn and Eugene Shoemaker and David Levy had not announced that it had broken up into twenty fragments – a cosmic 'string of pearls' moving through space.

Astrometrical calculations soon revealed that Comet Shoemaker–Levy 9 had fragmented because of tidal forces experienced during its previous passage near Jupiter in July 1992 – and all the fragments would fall inexorably into the atmosphere of Jupiter during the next encounter in July 1994. A mighty observational programme began, and numerous ground-based telescopes, working at all wavelengths, swung into action. In space, the Hubble Space Telescope, and the Galileo probe *en route* to Jupiter, trained their instruments on the comet.

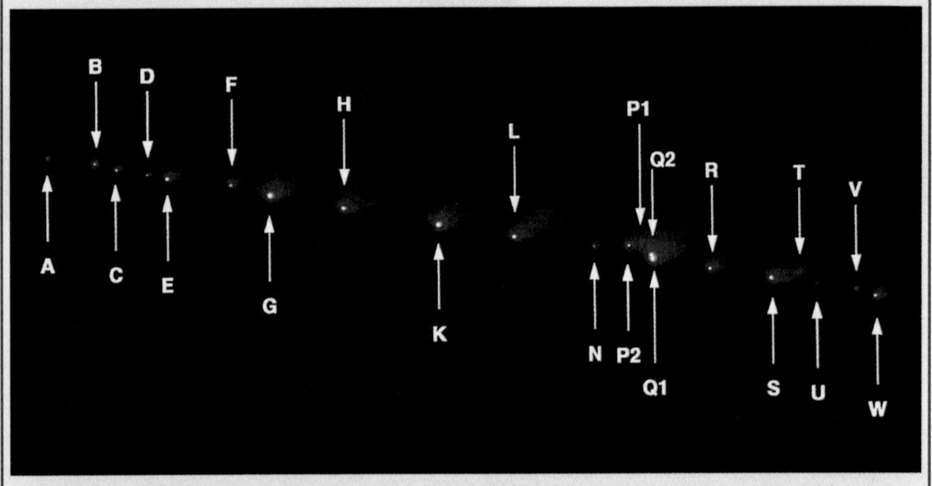

Figure 5.9. The various fragments of comet Shoemaker–Levy 9 were assigned letters from A to W. Some fragments (G, P and Q) then broke into two, and others (J and M) were volatilised. This image was taken with the Hubble Space Telescope's wide-field planetary camera (WFPC2) on 17 May 1994, two months before the collision. At this time the angular separation between the leading and trailing fragments A and W was 6 arcminutes, corresponding to a distance of just over 1 million km.

The series of collisions took place around Jupiter, at latitude 44° S, between 16 and 22 July 1994, within a few seconds of the times predicted by the astrometry. The impact sites appeared along this parallel as the planet rotated. The collisions resulted in tremendous explosions, at atmospheric levels equivalent to a pressure of approximately 0.2–2 bars. A ball of fire rose from each site, with matter ejected to an altitude (according to observations with the HST) of more than 3,000 km. Then, after about ten minutes, the ejecta fell back, heating the stratosphere to a temperature of more than 10,000 K and creating new molecules of H_2O, CO, HCN, CS, OCS, C_2H_4 and NH_3. Some of these chemical products were observed for several years after the event, their evolution proceeding in good agreement with photochemical models. The collision of comet Shoemaker–Levy 9 with Jupiter was an exceptionally important occurrence for astronomers. Statistics tell us that an event of this nature will happen every few hundred years. It may therefore be possible that it was a similar phenomenon that the astronomer Cassini observed at the end of the seventeenth century, judging by drawings in the archives of the Paris Observatory. The grandeur of the 1994 event showed us, in real time (and without danger to ourselves) how a planetary atmosphere reacts to a major meteoric impact.

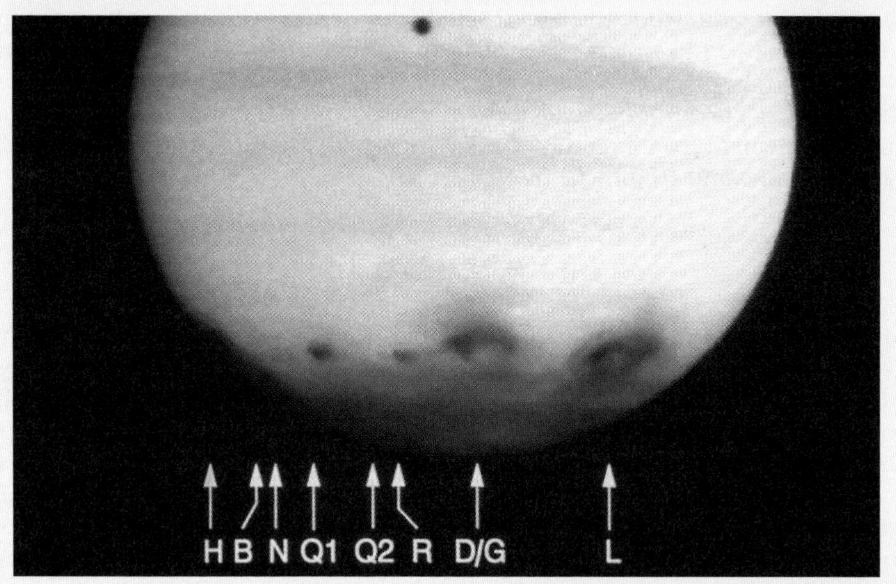

Figure 5.10. Various impact sites of some of the twenty-odd fragments of comet Shoemaker–Levy 9 that hit Jupiter in July 1994. The rotation of the planet caused the impact scars to be strung at different longitudes along latitude 44° S. This ultraviolet image was obtained by the Hubble Space Telescope, viewing the upper stratosphere of the planet. The collision was an exceptional phenomenon for astronomers, since, statistically, such an event occurs only once every few hundred years.

with sulphur dioxide and alive with volcanic activity – is nothing like Callisto, with its mass of craters unchanged for billions of years. Can we at least try to understand their overall chemical composition? There are two important factors here: the distance from the Sun, and the distance of the satellite from its primary.

The first factor is responsible for the composition of the ices in the planetesimals which, by accretion, formed the satellites. We have already seen that, at the distance of the giant planets from the Sun, the molecules most likely to be found in the solid state are H_2O, NH_3 and CH_4; and, at the distances at which Uranus and Neptune orbit, in a more rarefied environment, small quantities of CO and N_2 can also be present. This could explain the large amounts of CO and HCN in Neptune's atmosphere.

The second factor (the distance from the satellite to the planet) had some influence on the ambient temperature, since the central part of the sub-nebula is at a higher temperature than its exterior regions, in the manner of a solar system in miniature. There are other effects, too. The satellites are much nearer their primaries than the planets would be to the Sun in a scaled-down solar system. For example, Mercury, the nearest planet to the Sun, orbits at a distance of more than 80 solar radii, while the largest satellites are found at distances between just a few planetary radii and a few tens of radii. At distances like these, gravitational effects upon the satellites are considerable. Tidal effects – caused by differences in the planetary gravitational field at the near side and the far side of the satellite – are a supplementary source of internal energy for the inner satellites, and can cause volcanic and tectonic activity.

THE GALILEAN SATELLITES

After the collapse of the jovian sub-nebula, from which its system of satellites evolved, the temperature at the distance of Jupiter's orbit was certainly at a level where only H_2O, and a few minor constituents like SO_2 and CO_2 or certain salts, could condense. This is why, on the basis of current models, the structure of the Galilean satellites is likely to involve a silicate or metallic nucleus, surrounded by water, either as a liquid or in solid form as almost pure ice. This is indeed the case with the three outer Galileans: Europa, Ganymede and Callisto. As early as 1957, the American astronomer Gerard Kuiper had detected water ice on Ganymede and Callisto by means of near-infrared spectroscopy. However, Io, innermost of the Galileans, is an exception to the rule because the second factor (the tidal effect) comes into play, and temperatures vary strikingly with distance in the case of Jupiter.

Io

Orbiting at a distance of only six Jupiter radii, Io formed at a temperature sufficient for its water to escape. Only heavier elements remain: sulphur compounds, silicates, and metals. The density of Io is 3.5 g/cm^3, typical of silicates. Violent tidal effects involving Europa and Ganymede (the orbital parameters of the three satellites are affected by resonance phenomena) reinforce high temperatures and cause volcanism, as does the proximity of Jupiter. The Voyager, Galileo and Cassini missions provided a wealth of images of Io's volcanoes, and extended our knowledge of their physical properties and their evolution through time.

Figure 5.11. The four Galilean satellites. Current models suggest that these satellites should consist of a silicate and metal core surrounded by water – either liquid or as almost pure ice. This is the case with the three outer Galilean satellites, Europa, Ganymede and Callisto. However, Io, the nearest of these satellites to Jupiter, is an exception to the rule.

Europa

Europa, Ganymede and Callisto have silicate cores, surrounded by water in liquid or solid form. Europa, with a density of 3 g/cm^3, consists predominantly of the silicate core. However, what really interests scientists is its striated, highly reflective icy surface, beneath which may be water in viscous or even liquid form. What are the indications favouring this hypothesis? First of all there is the arrangement of the icy plates covering the surface, which appear to have drifted into position atop a more fluid, denser medium - and water is probably the carrier. Then there is the discovery by Galileo of an induced magnetic field, possibly generated internally by a salt-water ocean. Finally, there are the tidal effects due to both Jupiter and the resonances between the satellites. At nine Jupiter radii, these are less marked than in the case of Io, but they are no doubt strong enough to have prevented total condensation of the outer layers during Europa's cooling phase, and may preserve to this day an ocean of liquid water beneath the icy exterior.

If Europa has a hidden ocean, could it also harbour life? We cannot really answer this question without taking a close look - which is just the intention of planetary scientists with an interest in exobiology. The problem then will be one of scale. Assuming that this ocean is actually there, at what depth will it be found? Current (and very speculative) models of Europa's internal structure suggest that the thickness of the ice layer is several tens of kilometres - possibly even 100 km. The first step envisaged is an orbiter, programmed to detect this much-publicised ocean using gravimetric and altimetric methods. Next, perhaps, will be the *in situ* phase of exploration, when the drilling will begin.

Europa, with its possibility of extraterrestrial life, will doubtless be one of the prime targets of our planetary investigation during the decades to come.

Ganymede and Callisto

With a density of nearly 2 g/cm^3, Ganymede and Callisto are mixtures of water ice and rocky material. Ganymede, bigger than the planet Mercury, is the most imposing of all the satellites in the solar system. Its very heterogeneous surface has two main types of morphology: ancient, dark terrains which may be the remnants of its primordial surface; and lighter, somewhat younger zones laced with characteristic long grooves. Also, Galileo recently detected an intrinsic magnetic field, suggesting the presence of a molten, iron-rich core, probably of recent origin.

Callisto, at 26 Jupiter radii, is not in orbital resonance with the other Galileans, and therefore escapes the attention of tidal forces. With no internal energy supply, Callisto has a very densely cratered surface, unchanged for at least

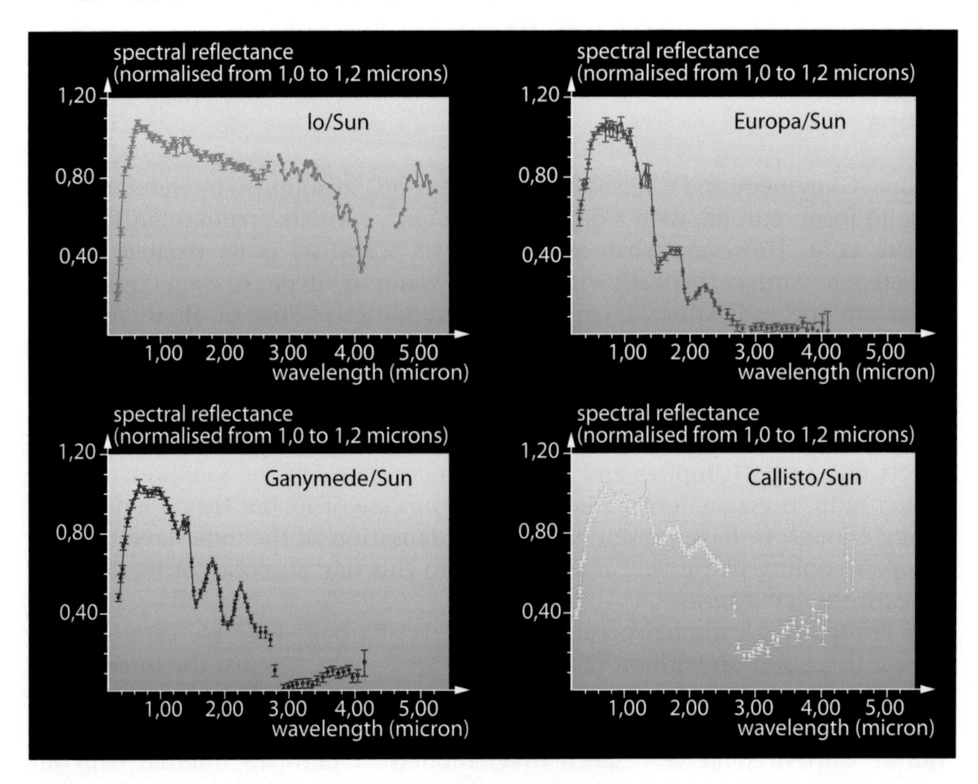

Figure 5.12. The spectra of the four Galilean satellites between 0.35 and 5 μm. The signature of water ice appears for Europa, Ganymede and Callisto at 1.5, 2.0 and 3.0 μm. However, the spectrum of Io shows the signature of SO_2 ice at 4.0 μm. (From T. Sill and R. N. Clark, *Satellites of Jupiter*, ed. D. Morrison, University of Arizona Press, 1982.)

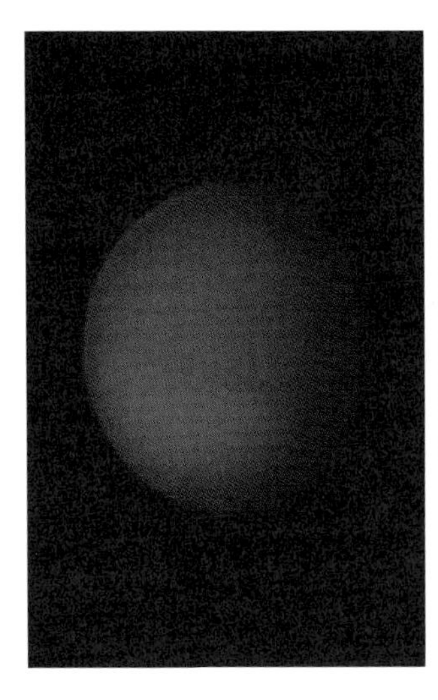

Figure 5.13. Titan, photographed by Voyager 1.

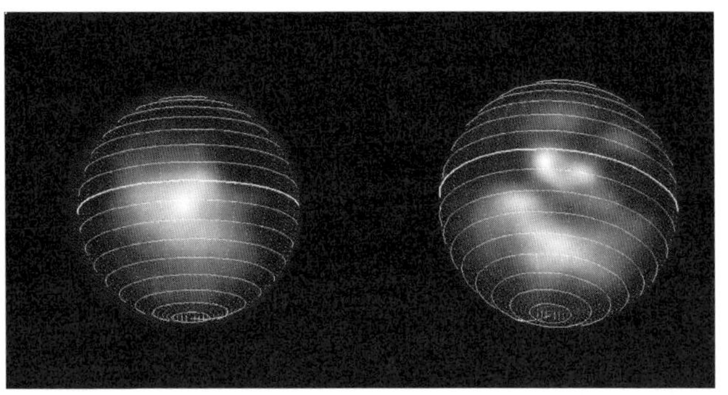

Figure 5.14. The surface of Titan observed from the Earth, using the technique of adaptive optics. The observations were carried out in the near-infrared, around 2 μm, providing a view through the cloudy blanket surrounding the satellite. (Left) the hemisphere observed by the Canada–France–Hawaii Telescope; (right) the other hemisphere, observed by the VLT (ESO). (Courtesy A. Coustenis.)

3 billion years. One giant crater, Valhalla, probably dates from an impact that occurred a few hundred million years after the formation of the globe. Thermochemical models of the internal structures of Ganymede and Callisto suggest that they too may have oceans within them.

SATURN'S SATELLITES

Titan: a laboratory for prebiotic chemistry

At 25 Saturn radii, Titan - the largest satellite of the saturnian system - closely resembles Ganymede and Callisto in size and density; but there the resemblance ends. Titan holds on to a dense, stable atmosphere, while the two Galileans have none. Titan's atmosphere is nitrogen-based which, with its surface pressure of 1.5 bar, gives it a slightly Earth-like quality. There is, however, a notable difference. Titan's surface temperature is only 93 K, and at the tropopause it is only 70 K. In 1980, Voyager 1 revealed the nature of Titan's atmosphere, its thermal structure and its clouds. Of considerable interest among its findings were a whole range of complex organic molecules, hydrocarbons and nitriles, detected using infrared spectroscopy. These are products of the photochemistry of methane and the dissociation of molecular nitrogen by energetic particles. These molecules are basic ingredients of the reactions which lead to the formation of amino acids, by way of mixtures containing H_2, CH_4 and NH_3. Titan would therefore appear to be an exceptional laboratory for prebiotic chemistry, potentially capable of adding to our knowledge of the first stages of the emergence of life. On 14 January 2005

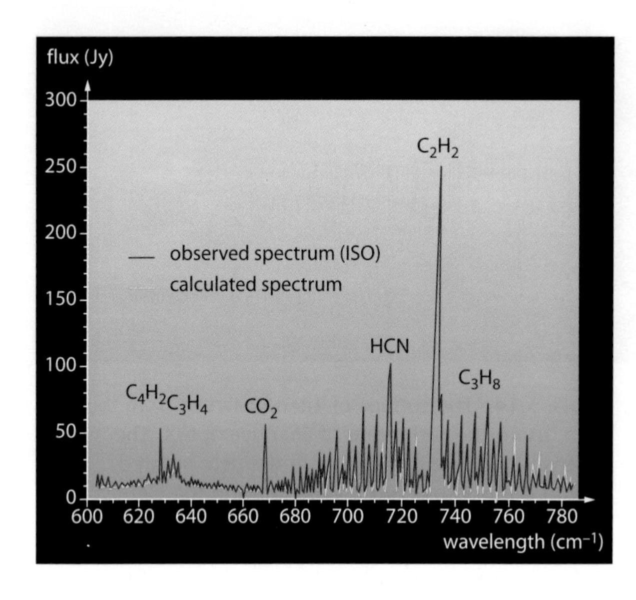

Figure 5.15. The infrared spectrum of Titan observed by the ISO/SWS in 1997 between 600 and 780 cm^{-1} (12.82–16.6 μm). (From A. Coustenis *et al.*, ESA SP-419, 255, 1997.)

a European probe, Huygens, descended through Titan's atmosphere and landed upon its surface. Its orbiting counterpart, Cassini, a joint ESA/ NASA project, is programmed to orbit Saturn and study the planet and its system.

Why does Titan possess a stable atmosphere, while Ganymede and Callisto have none? Similarities of size and density may not be matched by similarities in internal structure. At 20 Saturn radii, and 10 AU from the Sun, Titan formed at a considerably lower temperature. We would therefore expect the condensation of mixtures such as (H_2O, NH_3) or ($2H_2O$, NH_3), with densities close to that of water ice. According to theoretical models, apolar or weakly polar molecules (CH_4, CO, N_2, CO_2) could be fixed, either in the amorphous water ice or as clathrates. In clathrates, an atom or host molecule is trapped within a matrix of crystalline water ice. The present-day atmosphere of Titan – mostly N_2 with about 2% CH_4 by volume – could be explained by envisaging the entrapment of these molecules in the interior of the satellite. This would imply that Titan formed at a very low temperature. This hypothesis seems to be contradicted by the fact the mass spectrometer on Huygens did not detect argon-36, which also becomes entrapped at a very low temperature. Another, more plausible, hypothesis suggests that nitrogen was trapped within Titan as a component of NH_3 and not as N_2 – a situation compatible with the temperature expected in the saturnian sub-nebula. Outgassing of NH_3 would have led, by photodissociation of ammonia, to an atmosphere of N_2. The outgassing mechanism seems to be permanent, as the methane in the atmosphere is also destroyed by photo-dissociation and transformed into hydrocarbons. Its presence in the atmosphere implies the existence of an internal reservoir.

But where is Titan's water? Water ice is certainly present in the interior, but is it present at the surface? Images transmitted by the Huygens lander seem to indicate this (Figure 5.16). They reveal a fairly flat surface strewn with heavily

Figure 5.16. The surface of Titan, as revealed by the camera of the European Space Agency's Huygens probe, which landed on the satellite on 14 January 2005. The rounded 'pebbles' are probably made of water ice, and the ground appears to be covered with a deposit of hydrocarbons. The orange colour is due to the diffusion of sunlight by atmospheric aerosols rich in nitriles. (Copyright ESA.)

eroded 'pebbles', and various signs of the past activity of flowing liquid. It is probable that the erosion was caused by a fluid, now absent in the vicinity of the Huygens landing site, composed of hydrocarbons, and the 'pebbles' are quite probably composed of water ice. Infrared measurements of the surface support this interpretation. As for water vapour, we have seen that it has been detected in Titan's stratosphere (as well as in the stratospheres of the four giant planets). Here too, the existence of an external source of oxygen is the basis of a complex photochemistry, with the appearance of other types of molecule such as CO and CO_2.

Ice satellites

Within a distance of 10 planetary radii, Saturn has five large satellites in near-circular, equatorial orbits: Mimas, Enceladus, Tethys, Dione and Rhea. They increase in size with distance, and their densities – all close to 1 g/cm^3 – their high albedos and their infrared spectra imply a composition based on water ice. It is likely that the proximity of Saturn is the reason why the local temperature was high enough, at the time of their formation, for water alone to be in solid form. Their internal structure certainly comprises a small rocky core surrounded by a thick layer of ice.

The surfaces of these satellites are morphologically quite varied. Mimas, the nearest to Saturn, is very heavily cratered. Enceladus has one of the brightest surfaces known, indicating that the surface temperature is particularly low (70 K), as it absorbs only 10% of solar radiation. In 2005 the Cassini spacecraft imaged Enceladus, and to the scientists' surprise, revealed an active region at its south pole, where temperatures are locally higher than elsewhere and cryovolcanism is occurring. This phenomenon requires an as yet unexplained internal local energy source.

Like Jupiter, Saturn also has a large number of smaller moons, both close in, and far away from the planet; and more are discovered every year. Some of them are involved with the ring systems and play a part in their confinement; while others, towards the outside of the satellite systems, have high inclinations and very eccentric orbits, indicating that they have probably been captured.

THE COMPANIONS OF URANUS

Uranus has five main satellites, all orbiting within 25 planetary radii. In order of increasing distance from the planet, they are Miranda, Ariel, Umbriel, Titania and Oberon. They all orbit close to the equatorial plane of Uranus, and they must have formed at the time of the collapse of the gas surrounding the planetary nucleus, within the resulting sub-nebula. Their presence in the equatorial plane proves that they formed after the event (perhaps due to a collision, or the aftermath of some drastic phase of dynamic evolution) which toppled the axis of rotation of the protoplanet towards the plane of the ecliptic. Uranus' equatorial plane, with its family of satellites, is therefore nearly perpendicular to the ecliptic plane – a situation unique in the solar system.

The density of Uranus' satellites is about 1.6 g/cm^3. With an albedo of approximately 0.12, they appear much darker than the satellites of Saturn and Jupiter. Their infrared spectra show characteristic bands of water ice, less intense than those in the spectra of the Galilean satellites and Saturn's icy moons (Figure 5.17). A possible explanation for this is that their surfaces are composed not only

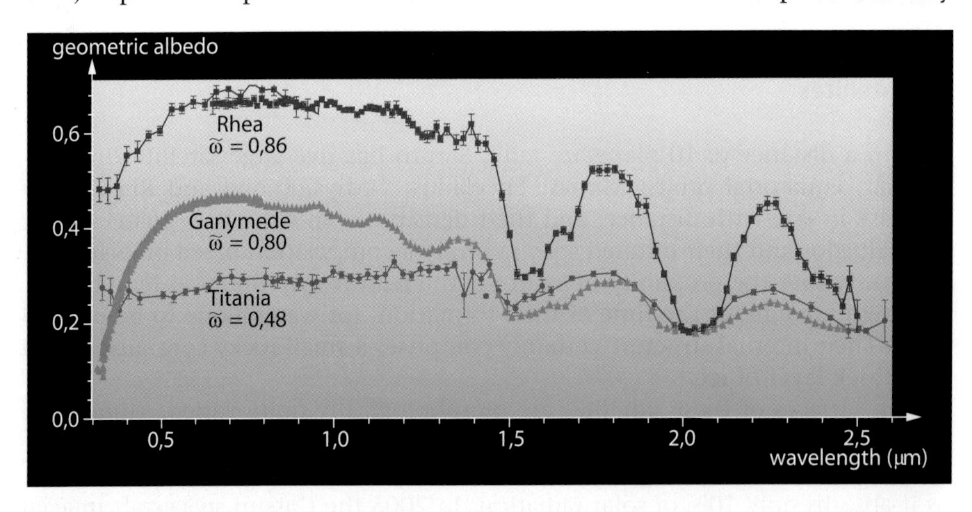

Figure 5.17. Spectrum of the reflectance of three ice satellites of three different planets. Saturn's satellite Rhea (in red) shows the most intense spectral signatures of water ice, at 1.5 μm and 2.0 μm. Even though the surfaces of the satellites are predominantly water ice, their reflectance is quite varied. (From Dale P. Cruikshank *et al.*, in *Solar System Ices*, eds. B. Schmitt *et al.*, Kluwer, 1998.)

of water ice, but also of other ices (NH_3, CH_4, and so on), and the dark material is the result of the irradiation of this mixture by cosmic radiation and energetic particles accelerated within the magnetosphere generated by Uranus' magnetic field. Laboratory experiments have shown that such irradiation will produce organic material which is dark and refractory, and similar results are possible in the case of cosmic radiation and intense ultraviolet radiation. The same effect is seen on the surfaces of comets and in the interstellar medium.

TRITON: AN EXAMPLE OF CRYOVOLCANISM

Next in our tour of the satellites of the outer solar system we come to Triton, a satellite of Neptune. Triton has received much attention from planetologists. Its main features, like those of Titan, were revealed to us by the Voyager mission. Before Voyager 2 flew by Triton in August 1989, we knew nothing of its surface

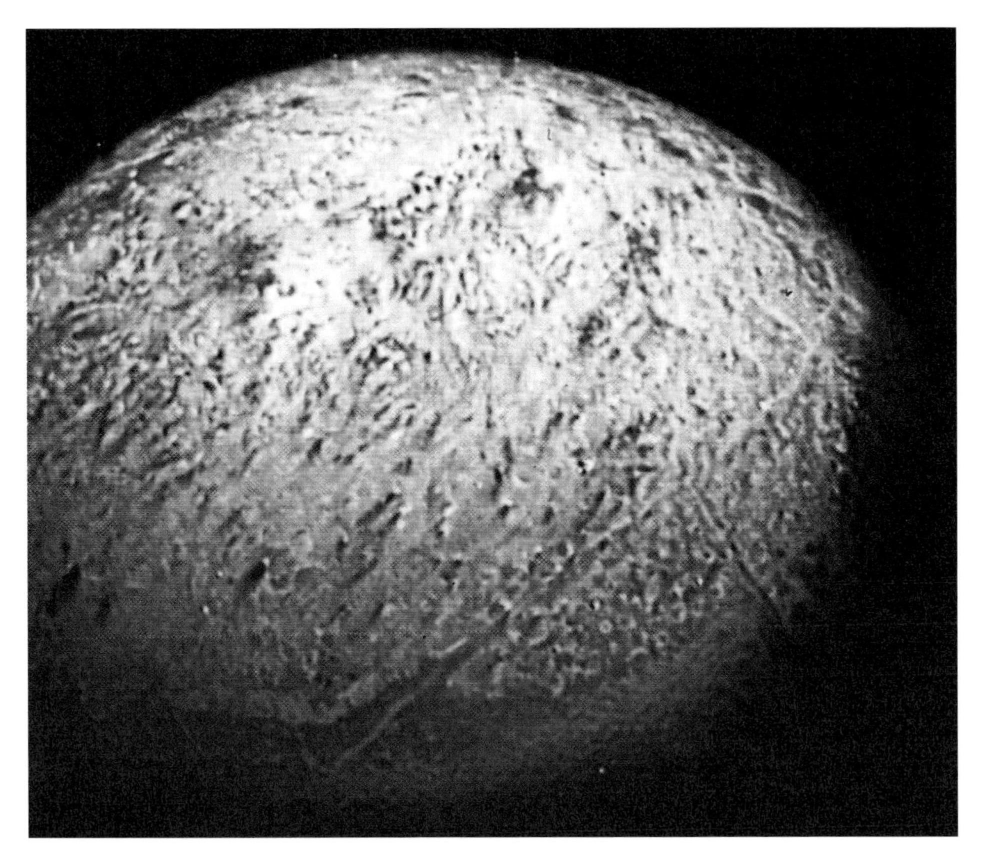

Figure 5.18. The surface of the southern hemisphere of Triton, imaged by Voyager 2. The dark streaks may well be caused by geysers or active volcanism.

pressure, its atmospheric composition or the nature of its surface. Again, Voyager 2's instruments exceeded all expectations.

Triton was discovered in 1846, a few weeks after Neptune was discovered. It orbits at nearly 15 Neptune radii, in an almost circular path which is both retrograde and highly inclined (157°), which suggests that it is a captured object. Due to its considerable inclination, in conjunction with that of Neptune, the seasonal effects are very marked, and unique in the solar system. The subsolar point (the point on the surface from which the Sun is at the zenith) migrates between latitudes +52° and –52°. Triton orbits Neptune in a state of captured rotation, always presenting the same face towards the planet, as do Earth's Moon, the Galilean satellites, and Titan. Before Voyager 2 made its discoveries, it was known that Triton's surface was composed of N_2 and CH_4 ices, detected by infrared spectroscopy; but nothing was known about its atmosphere or even its exact diameter and density. Voyager 2 measured the temperature of the surface at 38 K (the lowest ever in the solar system). Its albedo turned out to be quite high (0·7), and its surface pressure was 38 μbar. Its atmosphere is composed essentially of molecular nitrogen, with traces of methane (about 0.01%). The satellite can retain a stable atmosphere in spite of its modest dimensions (radius = 1,353 km), because its very low temperature slows the thermal motion of molecules and strongly inhibits their escape from the gravitational field.

Voyager images of the satellite's surface provided the greatest surprise. They show, in its southern hemisphere, a vast polar ice cap, to the north of which extends rough terrain which shows signs of past tectonic activity. Near the equator are more heavily cratered areas, which may date from 3 billion years ago.

The surface pressure on Triton indicates that the gaseous and solid phases of molecular nitrogen are in equilibrium. Triton's marked climatic effects induce a seasonal cycle involving the condensation of N_2 and CH_4, present in its ice caps. This is a similar situation to that on Mars, where the roles of N_2 and CH_4 are played by CO_2 and H_2O (to be discussed later). This cycle is marked by very violent winds, detected on the surface of Triton by Voyager 2. Images show dark streaks rising vertically and then moving off westwards to a distance of about 100 km. These plumes may well be a sign of active volcanism, or geysers activated by incident solar radiation, driven by a mechanism that remains unexplained.

How much do we know about the interior of Triton? Apart from measurements of its density and the identification of ice at the surface (N_2, CH_4, H_2O, CO and CO_2) (Figure 5.19), our ideas about its internal structure are based on theoretical models. It may have a core of metallic and rocky material, within a mantle of water ice, and above this, a layer of mixed ices containing water and other less refractory substances, CO, CO_2, NH_3, and CH_4 and N_2, which migrate from one polar cap to the other as the seasonal cycle proceeds.

What of the origin of Triton? It is very probable that it is a captured satellite. In a later section we shall see that it strikingly resembles Pluto, the ninth planet in the solar system, suggesting a common origin for the two bodies. We know nowadays that Pluto is a large member (and until recently, thought to be the largest) of a family of objects found in the outer reaches of the solar system: the

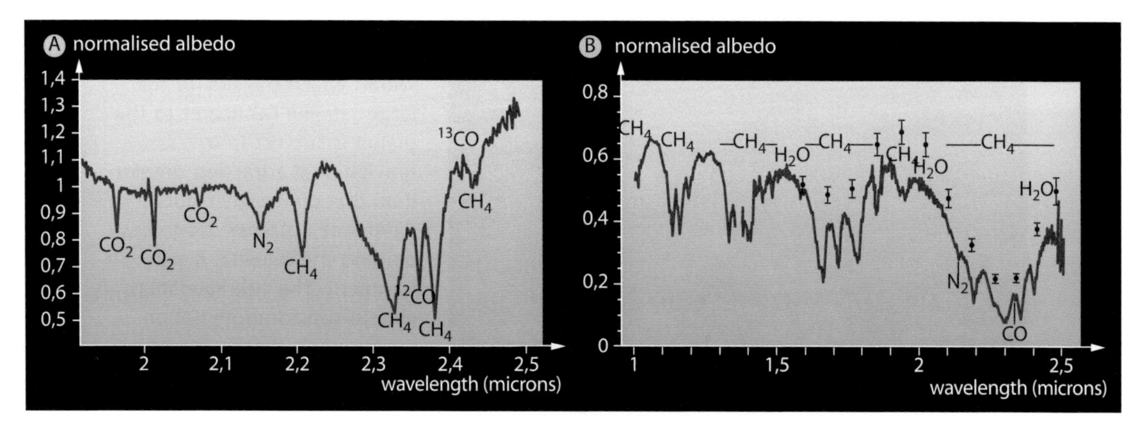

Figure 5.19. (a) The spectrum of Triton, observed from Earth in the near-infrared. The signatures of various types of ice are seen. Water ice is marginally present at around 2 μm. (b) The spectrum of Pluto in the near-infrared, showing the signatures of CH_4, N_2, and CO ices, and to a much lesser extent, H_2O. (From Dale P. Cruikshank *et al.*, *Ann. Rev. Earth Planet. Sci.*, **25**, 1997.)

trans-Neptunian or Kuiper Belt objects. Triton obviously belongs to this family, but is not well understood how it was captured by Neptune.

RINGS AND MINOR SATELLITES OF THE GIANT PLANETS

Having investigated the larger satellites of the giant planets, at distances of several radii from their parent bodies, what do we find as we move closer in? An important parameter comes into play here: the Roche limit (see p. 100). This is the distance within which a satellite is destroyed by differential tidal effects, on the sides turned towards and away from the planet, overcoming its cohesion. For a satellite of approximately the same density as its primary, this distance is about 2.5 times the planetary radius. In other words, within this limit we do not expect to find any stable satellites. This is the domain of the rings.

The rings of Saturn

It was long believed by astronomers that Saturn was the only ringed planet. Galileo first observed the rings in the early seventeenth century, and was puzzled by their changing appearance over time, because their inclination relative to Earth varies as Saturn orbits the Sun. Huygens explained this phenomenon for the first time in 1654, assuming that a (solid) disc lay in the equatorial plane of the planet. Not long afterwards, in 1675, Cassini discovered the division in the rings which bears his name. In 1785, Laplace showed that a solid ring could not remain intact given the tidal forces mentioned above. Later, Maxwell evolved the modern hypothesis that the rings are composed of individual small bodies, each

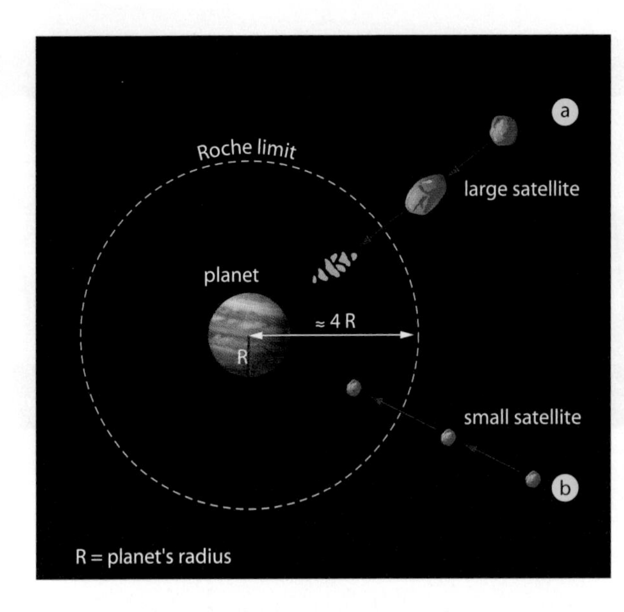

Figure 5.20. The Roche limit, within which the side of the large satellite (a) nearer to the planet is subject to a gravitational attraction greater than that on its other side. If this difference outranks the forces binding the satellite, it will fragment. The little satellite (b) is able to resist fragmentation because its cohesive forces are strong. The Roche limit is situated, according to circumstances, at a distance of 2–4 planetary radii from the centre of the planet, R representing its radius. (Modified from K. R. Lang and C. A. Whitney, 1993.)

Figure 5.21. The rings of Saturn, imaged by Voyager 1 on 12 November 1980.

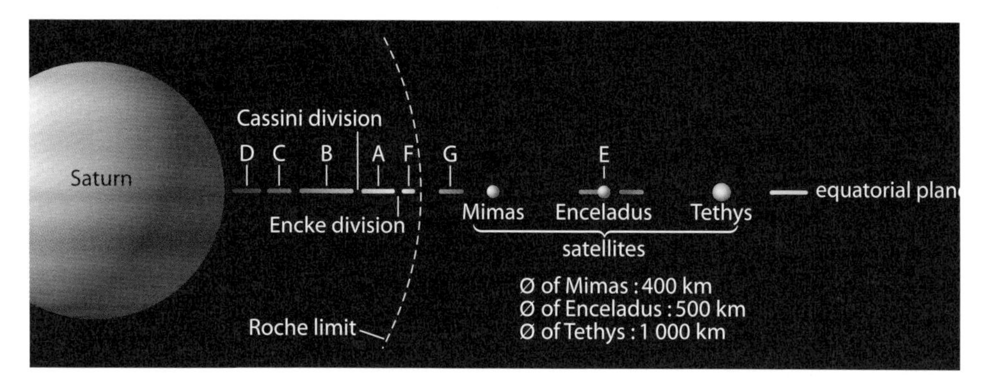

Figure 5.22. A cross-section of Saturn's rings and inner satellites. The Roche limit is situated between 3 and 4 radii from the centre of Saturn (R = 60,000 km). The principal rings, D, C, B, A and F, are situated within the Roche limit, and the satellites Mimas, Enceladus and Tethys are outside it. The G and E rings occur outside the limit, and five small satellites (of diameters 40–120 km) orbit near the Roche limit, within the orbit of Mimas. These small satellites are not shown on the diagram, which is not to scale. (Modified from K. R. Lang and C. A. Whitney, 1993.)

with its own orbital velocity. The spectacular images sent by Voyager 1 confirmed this theory.

One remarkable fact is that Saturn's rings are not totally confined to the space within the Roche limit: the E ring, on the outside of the system, extends out to three planetary radii, beyond the orbit of Mimas, and near the orbit of Enceladus. There are some small satellites orbiting within the Roche limit (Figure 5.22). These seem to play an important role in the configuration of the rings, and their lifetimes are limited. Although Saturn's rings are macroscopically stable on the scale of the age of the solar system, they display dynamical evolution on small time-scales, of the order of a few months or years.

What is the composition of Saturn's rings? Their infrared spectrum provides clear answers. Water ice is present in large quantities, and is confirmed by their high albedo and their low temperature (90 K). However, near-infrared spectra recorded by Cassini have revealed the existence of other constituents, and considerable variations in composition as a function of distance from Saturn.

Before the 1970s, Saturn's rings were the only planetary rings seen from Earth, showing the planet to be one of the most beautiful objects in the night sky. Since then, planetary ring systems have been discovered one after another, using two complementary methods: observation from Earth of occultations of stars by planets, and imaging by spacecraft.

The rings of Uranus

In 1977 an opportunity arose for astronomers to determine the temperature and mean density of Uranus' atmosphere as the planet passed in front of a bright star – a technique already tried for Jupiter (see fact box, p. 123) (Figure 5.24). Several

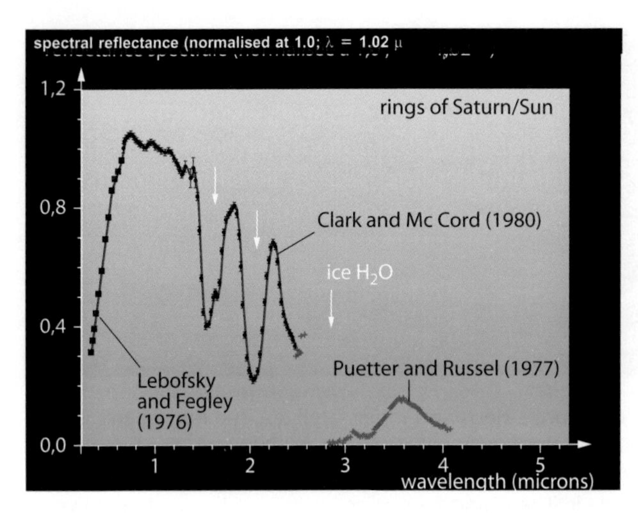

Figure 5.23. The spectral signatures of water ice in the rings of Saturn at 1.5, 2.0 and 3.0 μm. The absence of measurements around 2.8 μm is due to the presence of water vapour (see Chapter 1), which causes the Earth's atmosphere to be opaque. (From L. Esposito *et al.*, 1984.)

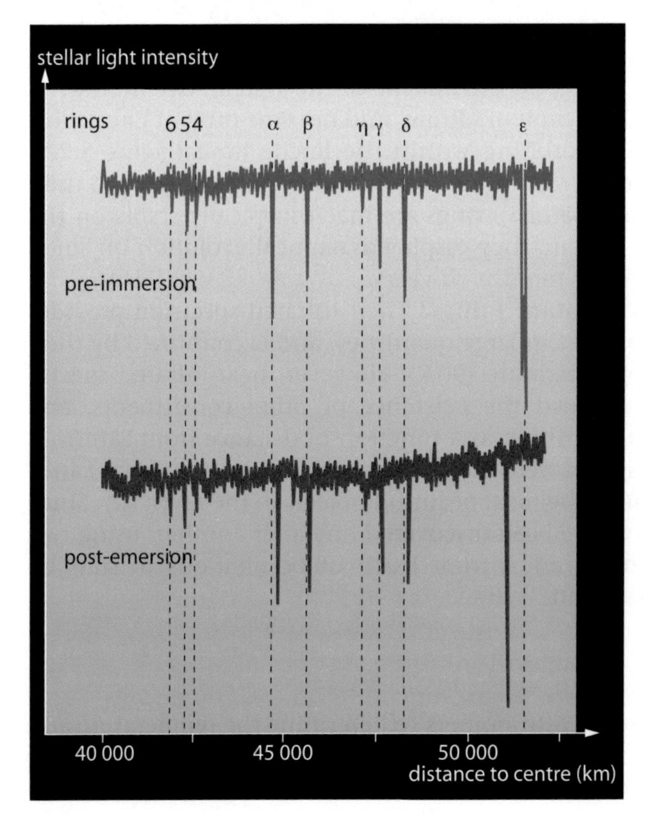

Figure 5.24. A light-curve of the occultation of the star SAO158687 by Uranus. This observation led to the discovery, on 10 March 1977, of Uranus' nine rings. These almost circular rings caused the apparent decreases in the brightness of the star. (From J. Elliot, *Ann. Rev. Astron. Astrophys.*, **17**, 445, 1979.)

telescopes, and the KAO, took part in this procedure, and on 10 March 1977 a *coup de théâtre* awaited the observers: measurements of the variations in the light of the star led to the detection of nine very narrow rings, as they too passed in front of it. These rings are located between 1.6 and 2 radii from the planet. It should be noted that in 1977 the south pole of Uranus was pointing in the general direction of both the Sun and the Earth, and the rings therefore appeared almost concentric. What struck the observers most was the extreme narrowness of these rings (less than 10 km across, for the most part) and their low albedo (0.05). In January 1986 the Voyager 2 probe flew past Uranus and provided an explanation. In In the case of the ε ring – at 36 km, the widest of them – its configuration is

Observing the giant planets and their systems via stellar occultations

When a planet passes in front of a bright star it is possible to glean information about the diameter of the planet by measuring the flux of the star as a function of time, if the precise position of the star is known through astrometry. If the planet has an atmosphere, immersion and emersion curves (plotting the decrease in the star's apparent brightness and its increase as the occultation finishes) will provide much information about the refractive index of the upper atmosphere. If the mean molecular mass is known (which is the case with the giant planets), then the temperature profile of the upper atmosphere can be deduced. If the planet possesses a ring system, its rings can also be detected by this method, as every time a ring or ring fragment passes in front of the star the star's brightness will abruptly (if briefly) appear to decrease. In this way, the nine narrow and tenuous rings of Uranus were discovered, by observers on Earth, on 10 March 1977. In 1984 this method was also used to detect Neptune's ring 'arcs', when the star dimmed briefly on one side of the planet but not on the other, suggesting the presence of an incomplete ring. In 1989, Voyager 2 confirmed the existence of complete rings, exhibiting maximum

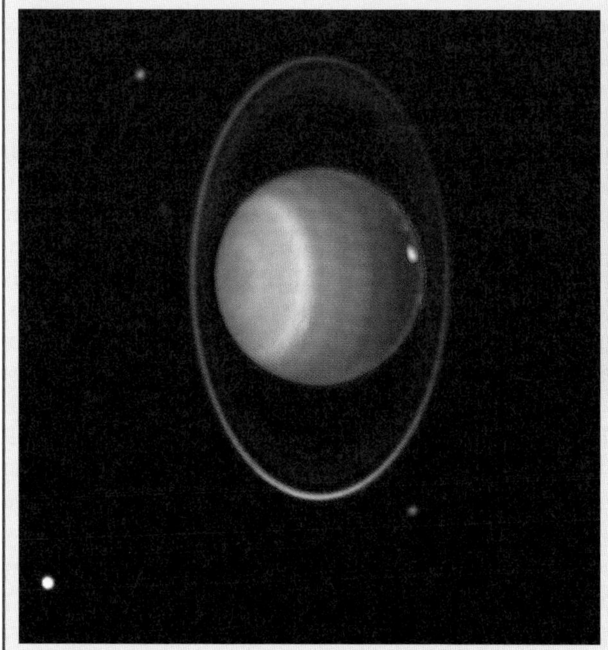

Figure 5.25. (a) The rings of Uranus, as seen by the Hubble Space Telescope in 1998. (Courtesy NASA/ESA). (b) The uranian system photographed in the infrared (2.2 μm) by the Isaac imaging spectrometer on the VLT in November 2002. The rings of Uranus are very tenuous, and would be almost invisible to the human eye. On the infrared image (b), the contrast between the rings and the planet is very exaggerated because, at this wavelength, sunlight reflected by the planet is almost completely absorbed by the methane in Uranus' atmosphere. (Photograph courtesy E. Lellouch *et al.*, ESO.)

density at certain points. These regions were the presumed 'arcs' detected during the occultation. Note that the occultation method (also known as the planetary transit method) is also very suitable for the detection of the passage of an extrasolar planet across the face of its star (see Chapter 9).

due to the presence of little 'shepherd' satellites moving on either side of it. The low albedo of the rings, like that of the satellites of Uranus, is doubtless due to the presence of organic polymers – products of the irradiation of ices by energetic particles in the magnetosphere.

The rings of Jupiter

The discovery of rings around Uranus was followed by the revelation, on 4 March 1979, of Jupiter's rings, detected by Voyager 1. At a distance of less than 2 Jupiter radii, these low-albedo rings are extremely tenuous and are undetectable from Earth. They are probably built up from grains escaping from the inner satellites of Jupiter: Thebe, Amalthea, Metis and Adrastea. These satellites are too close to Jupiter to have been able to conserve their water ice, and their composition is probably dominated, as is the case with Io, by silicates and sulphur compounds.

The rings of Neptune

Only Neptune's rings remained to be discovered, and once again the stellar occultation method provided the first observation of these, in 1984. But when they were detected there was a surprise in store. They were not rings, but ring arcs, visible on one side of the planet but not the other. A few years later, in August 1989, Voyager 2 confirmed the discovery and provided the key to the enigma. The probe found four narrow rings. One of them, afterwards named Adams, displayed three distinct bulges, corresponding to the arcs observed from Earth (Figure 5.26). *A priori*, such structures are not inherently stable, as the dust should diffuse in a short time along the length of the ring. Today, this phenomenon is thought to be due to a resonant interaction of mean movement 43:42 with the satellite Galatea, so that the ring performs 42 revolutions around the planet while the satellite performs 43 – a phenomenon already encountered in the case of the Galilean moons. The rings of Neptune, like those of Uranus, have a very low albedo, which suggests that their chemical composition is alike, with a mixture of ices coated with irradiated refractory material.

PLUTO AND THE TRANS-NEPTUNIAN OBJECTS

What lies beyond the orbit of Neptune? Astronomers have been asking this question for more than 100 years.

Figure 5.26. The rings of Neptune imaged by Voyager 2. Neptune's rings, like those of Uranus, have a low albedo, which suggests that they are similar in chemical composition (a mixture of ices covered in irradiated refractory material).

The two outermost giant planets, invisible to the naked eye, were discovered comparatively recently. In 1781, William Herschel discovered Uranus with a telescope of his own construction. In 1846 the discovery of Neptune, based on earlier work by John Couch Adams and Urbain Le Verrier, marked a great moment in the history of celestial mechanics. On the basis of their calculations involving anomalies observed in the orbital motion of Uranus, the two astronomers independently deduced the existence of Neptune and predicted its position in the sky; and then the astronomer Johann Galle found the planet.

However, the hunt for distant and unseen planets was not yet over. Calculations involving the motions of Uranus and Neptune revealed further anomalies. Astronomers explained these by invoking the presence of one or more massive planets at even greater distances. The search for these planets lasted for decades. In 1930, American astronomer Clyde Tombaugh discovered Pluto, which was pursuing a very eccentric orbit. But its mass - much less than that of the giant planets - was insufficient to explain the observed perturbations, and the hunt for a tenth planet continued until the end of the twentieth century. In 1992, another, smaller object, orbiting at 30 AU, was discovered by American astronomers David Jewitt and Jane Luu. This spectacular discovery owed nothing to chance. For five years, using ever more sensitive photometric methods evolved

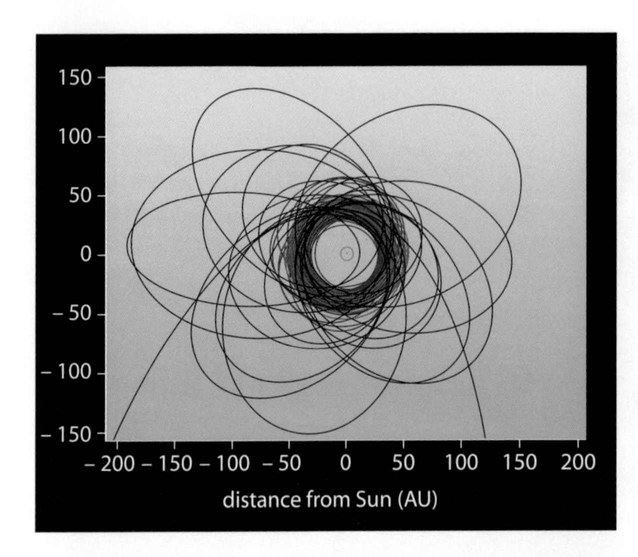

Figure 5.27. Paths of known trans-Neptunian objects on 1 January 2002, projected onto the plane of the ecliptic. (Courtesy D. Jewitt.)

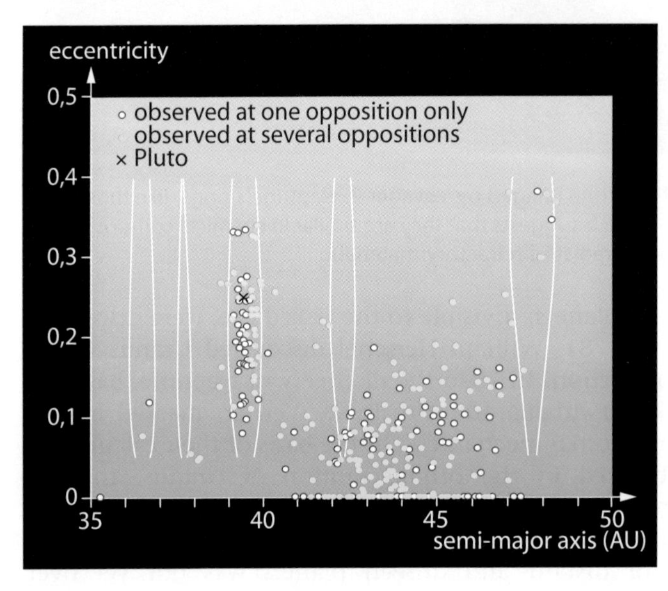

Figure 5.28 Eccentricities of trans-Neptunian objects as a function of semi-major axes of the orbits of objects discovered. Scattered objects are not represented here. The filled circles represent the positions of objects with well-established orbits. The various zones of resonance (4:3, 7:5, 3:2...) are indicated. Objects found at around 39 AU (such as Pluto) – the so-called Plutinos – are in a 3:2 resonance with Neptune; that is, their respective periods are in the ratio 3:2.

for use with large telescopes, the two researchers had carried out a systematic sky survey along the plane of the ecliptic. This was the first of a long series of observations leading to the discovery of a new class of object: the trans-Neptunians, orbiting in the Kuiper Belt. At the time of writing (January 2006) more than 1,000 trans-Neptunian objects are known, their diameters mostly ranging between tens of kilometres and several hundred kilometres. One of the largest yet discovered is 2003 UB313, with an estimated diameter greater than that of Pluto. The 'tenth planet' has never been discovered, and the supposed perturbations of Uranus and Neptune were no doubt artefacts of insufficiently accurate orbital data available in the past. The astronomers had been trying to solve a non-existent problem. However, the discovery of the Kuiper Belt represented a real revolution in planetary science, and pushed back the frontiers of the solar system.

It is clear that Pluto – the ninth planet, but difficult to classify among solar system objects – is one of the largest representatives of the trans-Neptunian family. Pluto is accompanied by a satellite, Charon, discovered in 1978 thanks to the improving quality of telescopic imaging. Charon orbits Pluto at a distance of 17 planetary radii, and its diameter is about half that of Pluto. Here is a veritable double planet, rather like the Earth and its Moon. It is worth mentioning that double systems have been detected in growing numbers among the population of trans-Neptunian objects.

What do we know about Pluto? First of all, it has a very eccentric orbit (e = 0·25) which causes very large variations in its atmosphere. When Pluto is at aphelion (which will happen in 2113), at 50 AU from the Sun, its surface temperature will be so low that its entire atmosphere will have condensed out. Pluto could be observed during the 1980s, at a distance of less than 30 AU, and it reached perihelion in 1989. What we know about Pluto is largely based on studies of mutual occultation phenomena of Pluto and Charon (between 1985 and 1990), and observations of a stellar occultation in 1988. Hubble Space Telescope images in the visible range, and infrared spectroscopy, contributed to the observation of these events. It was shown that the atmosphere of Pluto consists largely of N_2, with traces of CH_4, in equilibrium with the solid phase; the ices on Pluto's surface are of N_2, CH_4, H_2O and CO. Pluto bears a striking resemblance to Triton. It is tempting to imagine, therefore, that they have a common origin within the Kuiper Belt, although it is still little understood how Neptune captured Triton. Another unexplained phenomenon is the considerable lack of resemblance between Pluto and its satellite Charon. Charon's density (1.7 g/cm^3) is much less than that of Pluto (2 g/cm^3), its albedo is much higher (0·37 compared with 0·58), and the surface of Charon is mostly water ice. Given its surface temperature, it is unlikely that Charon has a stable atmosphere, and it is more like the satellites of Uranus than it is like Pluto.

New Horizons

NASA's New Horizons mission was launched in January 2006 – its objective being the *in situ* exploration of the outskirts of the solar system. It will encounter Pluto in 2015, and then go on to explore other trans-Neptunian objects between 2016 and 2020.

What is the nature of the trans-Neptunian objects? We have only hypotheses to guide us. We can deduce their diameters from their apparent brightnesses and thermal radiation. These objects are too faint to enable their surface composition to be investigated using infrared spectroscopy. However, astronomers have been able to measure their brightness in the visible. They appear to be a diverse group – some being neutral in colour (with a flat spectrum), and others somewhat reddish, suggesting a complex and heterogeneous composition. The neutral colour could be the result of a 'dirty ice' surface, while the red effect may be due to the presence of organic material produced by irradiation of the ice.

The Kuiper Belt may be viewed as the 4.5-billion-year-old fossil remnant of the protosolar cloud. It is not unreasonable to suppose that other planetary systems will also have such material in their outer reaches. Around some nearby stars – for example, ε Eridani – discs of dust with dimensions comparable to those of the Kuiper Belt have been detected. An excess of water vapour around the star W Hydrae could be evidence of the volatilisation of its 'Kuiper Belt', as the star heats up during its transformation into a red giant. The study of the Kuiper Belt can therefore provide vital information about the formation of planetary systems.

6
At the ice line: the asteroids

At about 3 AU from the Sun, between Mars and Jupiter, thousands of asteroids pursue their orbits, which are sometimes strongly elliptical. If the Earth draws them in with its gravitational field, and they penetrate its atmosphere, any bodies weighing less than 1 kg will be abraded and volatilised. These are the meteors (shooting stars). Is there water in the asteroids? Those in the inner solar system have none, and detecting water in the others is not easy. It could well be that certain asteroids are merely comets which have lost all their ice through sublimation, during many visits to the neighbourhood of the Sun.

MINOR PLANETS

Leaving the outer solar system, a realm of ice, we move Sunwards and explore the inner solar system, dominated by the terrestrial planets. On the way we encounter a zone of transition: the main asteroid belt. This region, situated between Mars and Jupiter at about 3 AU from the Sun, is the home of the asteroids, or minor planets.

In 1801, Sicilian astronomer Giuseppe Piazzi discovered the first and largest object in this family: Ceres, with a diameter of just under 1,000 km. Not long afterwards, three other objects, smaller than Ceres, were discovered pursuing similar orbits: Pallas, Juno and Vesta. Others followed as telescopic techniques improved. There are now more than 100,000 known asteroids, and more than 20,000 of them have accurately determined orbits. Most of them travel within the main asteroid belt, between 2 and 3.5 AU from the Sun. Asteroids follow paths which are sometimes strongly elliptical, with eccentricities between 0·01 and 0·3. Their orbits are not much inclined (at angles less than 30°) relative to the plane of the ecliptic (Figure 6.2).

Figure 6.1. This meteoritic fragment, which fell in Western Australia in 1960, may be part of the asteroid Vesta. Like Vesta, it is made exclusively of pyroxene, a crystallised silicate.

A shower of meteors.

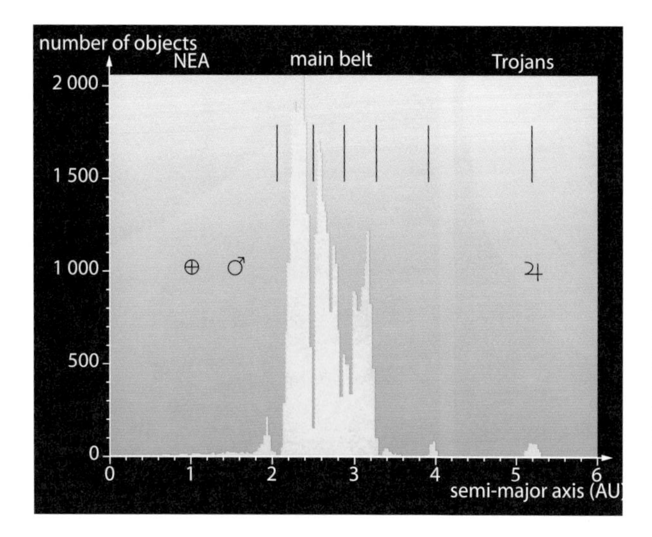

Figure 6.2. Distribution of the semi-major axes of asteroids. Most known asteroids follow orbits which are highly elliptical, with eccentricities between 0.01 and 0.3, and often only slightly inclined (at less than 30°) to the ecliptic. The symbols for Earth, Mars and Jupiter show the positions of their orbits.

What is the origin of these minor planets? One early theory held that they are the aftermath of the destruction of a large planet orbiting between Mars and Jupiter. Today, in the light of modern accretion scenarios, astronomers have adopted the different explanation that the asteroids formed locally from planetesimals present between Mars and Jupiter. The planetesimals were unable to form a large planet because of perturbations caused by Jupiter, and the effect of its powerful gravitational field was the dispersal of most of the planetesimals. Those that escaped this process were accreted into a family of small objects with stable orbits.

As well as the asteroids of the main belt, other families of objects travel within the solar system, on either side of the main reservoir. On our side of the asteroid belt travel the 'Earth grazers', whose orbits pass close to Earth's orbit, and some of which have collided with our planet in the past. It is now commonly believed that one such large object, about 10 km in diameter, was responsible (at least in part) for the extinction of the dinosaurs and many other species 65 million years ago (see fact box, p. 134). Keeping watch for these Earth-grazers, and studying their nature, has become a priority within programmes assessing natural hazards.

Outside the main belt are found the Trojan and Greek families of asteroids, moving along the orbit of Jupiter at an angle of 60° respectively in front of and behind the planet at the Lagrangian points, which offer a mechanically stable configuration. Even further from the Sun, the Centaurs orbit between Jupiter and Neptune.

The chemical and mineralogical composition of the asteroids is well known, due to spectroscopic observations carried out on Earth in the near-infrared. These measurements have been complemented by observations of certain asteroids by passing spacecraft: Gaspra, and Ida and its satellite Dactyl by the Galileo probe; Mathilda and Eros by NEAR; and Braille by Deep Space 1. All these observations have led to the classification of asteroids into several main categories, the

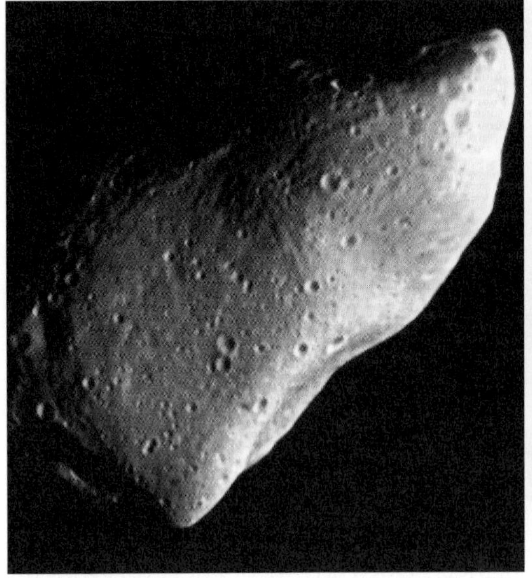

Figure 6.3. The asteroid Gaspra imaged by the Galileo probe as it flew by in October 1991 at a distance of 5,300 km. Gaspra measures approximately 19 × 11 km. (Courtesy NASA.)

Figure 6.4. A Galileo image of the asteroid Ida from a distance of 10,900 km. Ida is about 56 km long. It has a satellite Dactyl, also discovered by the Galileo probe. (Courtesy NASA.)

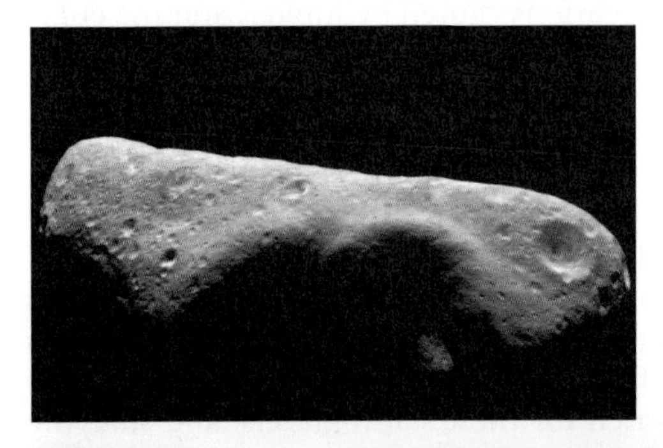

Figure 6.5. The southern hemisphere of the asteroid Eros, photographed from the NEAR Shoemaker probe on 30 November 2000. Eros is about 33 km long. (Courtesy NASA.)

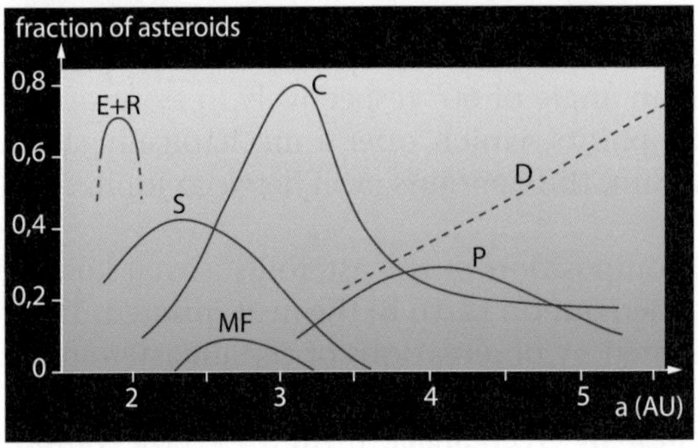

Figure 6.6. Spectral types of asteroids as a function of their heliocentric distance. Note especially that S-type (metal-silicate) asteroids are predominant in the inner region of the main belt, while C-types (carbonaceous and rich in hydrated silicates) are the most common in the outer part. Only the most primitive (D-type) asteroids are present at great heliocentric distances. (From T. Encrenaz *et al.*, *Le Système Solaire*, CNRS-Editions/EDP-Sciences, 2003.)

principal ones being metallic (M-type), silicaceous (S-type) and carbonaceous (C-type). It is interesting to note that the chemical composition of asteroids varies according to their distance from the Sun (Figure 6.6). Within the inner solar system are metallic and silicaceous objects, formed at a relatively high temperature, while the carbonaceous objects are found mainly within the principal asteroid belt. The D-type objects, characterised by surfaces with low albedo and a rather red appearance, due to irradiation, are the most primitive of all. These are found near the orbit of Jupiter and beyond.

Meteor showers of cometary origin

Figure 6.7. Meteors ('shooting stars') are micrometeoroids heated in the Earth's atmosphere and volatilised before reaching the ground. Their luminous trails all seem to emanate from the same point in the sky – the radiant – which marks the direction from which the meteor swarm is coming.

Meteors are the visible trails of micrometeoroids, heated by the Earth's atmosphere and vapourised before they can reach the ground. The number of meteors seen in the sky is not constant, and there are strong maxima at certain times of the year. During these showers, the luminous trails of the meteors all seem to originate from the same point in the sky – the radiant, named after the constellation in which it lies.

Where do these showers originate? They are part of the swarms of cometary debris scattered along the paths of periodic comets. When the Earth's orbit intersects these paths, its gravity draws in large numbers of dust particles which all seem to emanate from the same spot, being the point where the Earth's orbit meets that of the comet.

The principal meteor displays are the Aquarids (1–8 June) and the Orionids (18–26 October), associated with Halley's comet; the Perseids (25 July–17 August), associated with comet Swift–Tuttle; the Draconids (9–10 October), associated with comet Giacobini–Zinner; and the Leonids (14–21 November), associated with comet Tempel–Tuttle.

ASTEROID OR COMET?

Is there water in asteroids? As far as those in the inner solar system are concerned, the answer is no. C-type objects, associated with the main belt, show evidence of modification by water - a result of the possible entrapment of water in the liquid state by planetesimals originally moving at this distance from the Sun. As for the primitive D-type objects, they probably have a proportion of volatile elements, but irradiation of their surfaces has made detection of ices

The Chicxulub crater and the K-T boundary

The Chicxulub crater, identified in the 1990s, is an impact feature about 170 km across, situated on the northern coast of the Yucatan peninsula in Mexico. It is not evident at the surface, having been covered by sediments about 1 km thick, but its presence was suggested by satellite images showing arc-shaped structures on gravimetric charts. Later, micrometeorites (small glass beads formed from material melted at the moment of impact) were identified, and the age of the crater was found to be 65 million years.

The size of the crater suggests that the diameter of the impactor was around 10 km. The energy liberated by this impact may have attained a level of a hundred times that produced by one of the collisions of comet Shoemaker–Levy 9 with Jupiter, equivalent to a billion hydrogen bombs. The consequences of such an event must have been considerable for the environment. The amount of ejected material suspended in the atmosphere for several years would have seriously affected the climate by blocking out sunlight.

It is probable that the Chicxulub impact was responsible, at least in good measure, for the mass extinctions which occurred on Earth 65 million years ago. The signature of these extinctions appears in the sedimentary rocks of that epoch (at a point called the Cretaceous–Tertiary (K-T) boundary), in the form of a sedimentary layer rich in iridium. Fossils are present in the deeper, older layers, but absent in those above. The iridium layer probably came from the material of the asteroid, as iridium is normally absent from the Earth's crust, having been trapped in its iron core, though another theory suggests that some of it may be volcanic in origin.

It is thought that an asteroid more than 1 km wide will hit the Earth about once every 100 million years. Those engaged in the scientific study of natural hazards are turning their minds to possible ways of preventing similar events in the future. The first task is to create an inventory of all asteroids with orbits that bring them close to the Earth (near-Earth objects – NEOs), and the second will be to explore how spacecraft technology might be used to divert any object likely to be a danger.

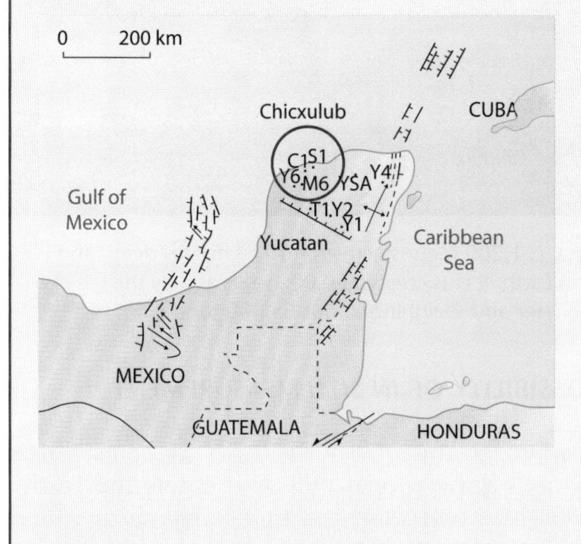

Figure 6.8. The Chicxulub crater, 180 km across, straddles the north-western coast of the Yucatan peninsula. Buried beneath 1,000 metres of chalk, half of its basin lies beneath the Gulf of Mexico, the other half beneath the land. At the surface its shape can be deduced due to the alignment of water features, influenced by subterranean rifts. Sites where drilling has taken place are indicated by letter symbols (C1 = Chicxulub 1, S1 = Sacapuc 1, and Y1 = Yucatan 1). (From C. Frankel, *La Mort des Dinosaures*, Masson, 1999; and K.O. Pope *et al.*, 1993 (modified).)

difficult. This is also the case with the Trojans and the Centaurs. For example, the orbit of Chiron, the largest representative of the Centaur family, extends as far out as the orbit of Uranus, and as far inwards as the orbit of Saturn. The diameter of Chiron is 180 km. Surprisingly, photometric variations observed in visible light have revealed possible outgassing on this object. Is Chiron an asteroid, or is it a giant comet? Indeed, the distinction between the two types seems to be increasingly blurred. Chiron is undoubtedly one more element in that class of trans-Neptunian objects described above. Moreover, there are more and more examples of asteroids which have been discovered following the orbits of former comets. It is likely that they have lost all their ice, after many perihelion passages, and there is now no difference between them and the carbonaceous asteroids. There is seemingly little but a step in time between comets and asteroids.

Figure 6.9. Meteor Crater, Arizona, is 1,200 metres across and 175 metres deep, and is one of the largest known craters on Earth. It was created 50,000 years ago by the fall of a meteorite about 25 metres in diameter and weighing about 65,000 tonnes.

METEORITES: THE POSSIBILITY OF *IN SITU* MEASUREMENT

Certain asteroids are known to approach the orbit of the Earth. Much asteroidal material is captured by our planet's gravitational pull, and enters the Earth's atmosphere, where abrasion volatilises objects of less than 1 kg and produces meteors. The displays of meteors, which appear on certain dates, are associated with the passage of the Earth near the orbits of periodic comets (Halley, Swift–Tuttle, Giacobini–Zinner, and so on). They all seem to be of cometary origin, as are the interplanetary dust grains (micrometeorites) found on Earth.

Larger objects may fall as meteorites. The largest known example is the Hoba meteorite, found in Namibia, which weighs 60 tonnes. The largest meteorite crater to be identified is the Chicxulub feature in Mexico, created 65 million years ago by the devastating impact responsible for the ecological catastrophe marking the transition between the Cretaceous and Tertiary periods (see fact box, p. 134).

Leaving aside the potential but fortunately very improbable threat of such objects colliding with Earth, (almost) without warning, the study of meteorites is very important in our understanding of the history of the solar system.

Indeed, the abundances of the elements found in the most primitive meteorites – with the exception of the most volatile elements, which could

Heavy water in meteorites: a clue to the formation of the solar system

The D:H ratio is a quantity which depends on the chemical reactions involved in the origin of the water molecule in different solar system objects (planetesimals, comets, asteroids and planets). Within a cold, dilute environment suffused with ultraviolet radiation from nearby stars, ion–molecule and molecule–molecule reactions favour excess of molecules other than hydrogen (which is the main reservoir of deuterium in the form of HD). This is particularly the case in the reaction:

$$H_2O + HD \rightleftarrows HDO + H_2$$

which leads to an excess of HDO at low temperatures. This HDO/H_2O excess can be measured as ratio *f* defined as follows:

$$f = (D/H)_{H_2O} \,/\, (D/H)_{H_2}$$

In the protosolar nebula, $D/H = 2.5 \times 10^{-5}$; and in the case of the Earth, the ratio *f*, measured in the oceans, is equal to 6($D/H = 1.5 \times 10^{-4}$).

In the interstellar medium, at a temperature of the order of 20 K, the deuterium excess in certain molecules such as HCN or HNC may attain very high values (for example, 2,000). The fragmentation of clouds like these led to the collapse of the protosolar nebula (see Chapter 3). By measuring factor *f* in the various objects of the solar system, we can obtain an indication of the temperature at which they formed, and of their distance from the Sun at the time. Theoretical models exist which enable us to quantify this effect. They predict that after a few million years, the nebula having cooled, the equilibrium value for *f*, within a distance of 3 AU from the Sun, will be 3.5.

Laboratory analysis of meteorites contributes greatly to these studies. Two types of meteorite contain water in the form of clays: carbonaceous chondrites and LL3 chondrites, characterised by a low iron content. While measurements of *f* in the carbonaceous chondrites show a weak dispersion around $f = 6$, the LL3 chondrites produce a wide range of values from 3.5 to 29. This shows that LL3 chondrites have preserved the isotopic heterogeneity of the water in the primordial nebula by conserving intact the ice grains formed at different temperatures and heliocentric distances. The maximum observed value bears the signature of the interstellar medium, while the minimum value seems to correspond with the equilibrium vale attained after the cooling of the nebula. According to this scenario, it might be expected that the value of *f* in comets would be high, since they formed far from the Sun at low temperatures. And indeed, measurements of Halley's comet in 1986 bore this out, as did later observations of comets Hyakutake and Hale–Bopp. In all three cases the D:H ratio was 3×10^{-4}, or $f = 12$ (see Chapter 4).

All these findings offer clues as to the origin of the Earth's water. At 1 AU from the Sun, the temperature of the nascent Earth was too high to lead to the excess (of value 6) as observed in its oceans. An external cause is therefore necessary, involving planetoids and comets formed at great distances from the Sun (see Chapter 7).

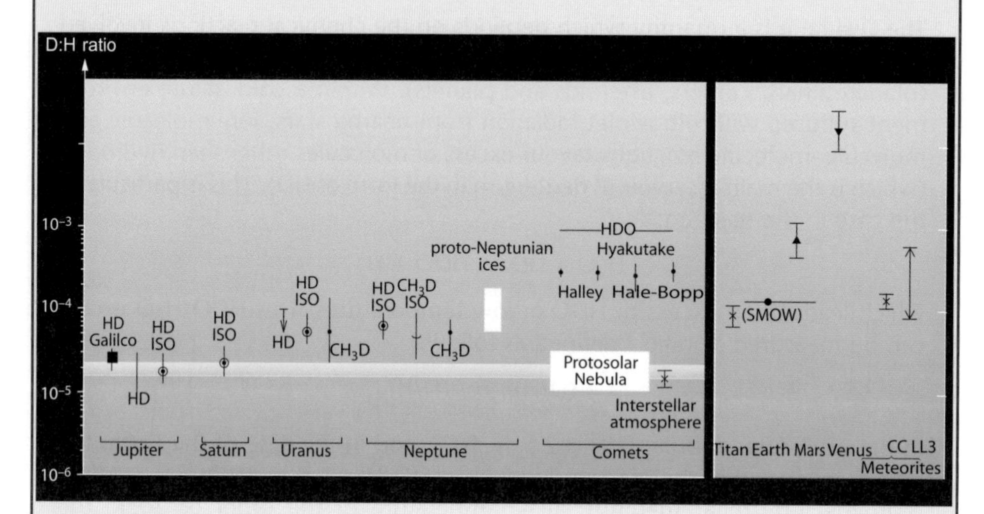

Figure 6.10. The D:H ratio in the solar system. In the case of the giant planets and Titan, the D:H ratio is based on measurements of $HD:H_2$, or $CH_3D:CH_4$. In the case of comets and terrestrial planets it is based on measurements of $HDO:H_2O$. By measuring D:H for various solar system objects we can find clues to their temperatures and heliocentric distances at the time of formation. In the ice giants Uranus and Neptune, a D:H excess is observed relative to the protosolar D:H value, in agreement with the 'nucleation' model of planetary formation (see Chapter 3). It is also noteworthy that the D:H value in comets is higher than the terrestrial value, as measured in the oceans (SMOW – Standard Mean Ocean Water). This shows that the water of Earth's oceans cannot have come entirely from comets, and must have come mainly from carbonaceous chondrites with a D:H value close to the terrestrial value. Finally, the considerable deuterium excesses measured in the water of the atmospheres of Mars and especially Venus are doubtless the result of differential escape rates (heavy water HDO escaping less readily than H_2O). They are also the signature of massive outgassing during the early history of the two planets.

not be fixed within planetesimals – are in close agreement, and provide accurate dating of the origin of our solar system. Its age has in this way been determined to be 4.55 billion years, to an accuracy of the order of 1% (see Chapter 3). The analysis of meteorites has also provided us with information about the composition of asteroids, since in many cases it is possible to associate a meteorite with a parent body belonging to one of the known families of asteroids. Carbonaceous chondrites, for example, have a chemical and mineralogical composition very similar to that of the carbonaceous (C-type) asteroids. In this context, the measurement of the D:H ratio takes on a particular interest (see fact box, p. 137).

Lastly, it should be mentioned that at the time of writing (January 2006) more than twenty known meteorites are thought to be of martian origin. We shall see below how they have contributed invaluable new information to complement measurements of chemical composition carried out by Mars probes. Laboratory analysis, employing the most technically efficient methods, has brought results far more accurate than those gained by spacecraft.

7
Water and the terrestrial planets

Mercury, Mars, Venus and the Earth, and their satellites, having formed nearer to the Sun, have evolved in very different ways. Water is very abundant on the Earth and in its atmosphere, but on Mercury it is present only as ice at the poles, on Venus as vapour in the atmosphere, and on Mars as ice at the poles and vapour in the atmosphere. Why? The principal explanation must be sought in phase changes of the H_2O molecule with temperature.

Figure 7.1. Mercury, as seen by Mariner 10 on 24 March 1974. Saturated with meteorite craters, the surface of Mercury resembles that of the Moon.

Formed near the Sun and composed mostly of silicates and metals, the four terrestrial planets are characteristically small and of high density (between 3.9 and 5.5 g/cm^3). Their atmospheres represent only a small fraction of their masses, and in the case of Mercury, practically none. They have few satellites, or none, and lack ring systems. In Chapter 3 we saw that these general properties fall naturally into the model by which planets form by accretion, from planetesimals, within a protoplanetary disc. If we take a closer look at these planets and their (few) satellites; we immediately see that they are all very different from each other.

Has there ever been liquid water in the craters of Mars? This artist's impression shows the possible appearance of the crater Fesenkov 3.8 billion years ago.

MERCURY AND THE MOON: NO ATMOSPHERE, BUT TRACES OF WATER?

Mercury

Mercury is the nearest planet to the Sun, and the smallest of the terrestrial planets. With a radius of slightly less than 2,500 km, it is smaller than even Ganymede and Titan, moons of the outer solar system. It lacks a stable atmosphere – which is easily understood, since the thermal motion on Mercury of any gas molecules, which is proportional to the temperature, is greater than the planet's escape velocity (4.4 km/s). (The smaller the planet, the lower its escape velocity.) It is very difficult to observe the surface of Mercury from the Earth, because the planet is never more than 30° away from the Sun in the sky. In 1974 the American probe Mariner 10 made three passes of Mercury and revealed the planet to us. Peppered with impact craters, Mercury might easily be mistaken for the Moon. Both of these worlds, unprotected by any atmosphere, bore the full

Figure 7.2. Like Mercury, the Moon lacks a stable atmosphere. Its heavily cratered surface is composed mainly of silicon oxides rich in aluminium and calcium. Very small amounts of water ice may be trapped in the vicinity of the poles, in permanently shadowed areas.

brunt of the very intense meteoritic bombardment which raged for a billion years after the formation of the planets. Mercury's surface appears to be made mostly of silicates, and iron and titanium oxides.

Nobody expected that any signs of water would be detected on the planet, but in 1991 radar images of its surface, at high angular resolution, revealed areas of increased reflectivity and strong polarisation at its poles, characteristic of water ice. Mariner 10 images had shown that there existed, near the poles, areas where crater interiors are in permanent shadow. Since Mercury has almost no axial tilt relative to the ecliptic, it has no seasons, and the temperature in these polar regions must be below 135 K. It is therefore plausible that a thin film of water delivered by meteorites, and especially comets, impacting on Mercury, might accumulate in these areas. The planned Messenger and BepiColombo missions of the next decade will be able to investigate this hypothesis.

The Moon

Like Mercury, the Moon, with an escape velocity of 2.3 km/s, has no stable atmosphere. It is locked in a captured rotation as it circles the Earth, and the two bodies form a kind of double planet with an origin long shrouded in mystery. Nowadays, the generally accepted theory is that the Earth–Moon system was formed about 4.5 billion years ago, when a protoplanet the size of Mars collided with the young Earth. This impact involved both the ejection into orbit around the Earth of part of the mantles of both bodies, and the fusion of two cores rich in heavy elements, after which the orbiting debris reaccreted to form the Moon. This scenario goes a long way towards explaining the odd fact that the density of the Moon (3.3 g/cm^3) is much less than that of Earth (5.5 g/cm^3).

The Moon is undoubtedly the most intensely scrutinised of all heavenly bodies, both telescopically and more directly by *in situ* analysis of surface material in the 1960s and 1970s. It is dominated by ancient and mountainous landscapes, heavily cratered, and the surface composition is mostly silicon oxides rich in aluminium and calcium. The remainder of its terrain consists of basaltic plains (the maria, once erroneously thought to be seas), produced by lava outflows from the mantle. Is there water ice on the lunar surface? The bold vision of Hergé, in the Tintin adventure *On a Marché sur la Lune* (*Explorers on the Moon*) may no longer be valid, but this in no way detracts from the genius of the visionary cartoonist.

However, there may be water near the Moon's poles, as is the case with Mercury. The American Clementine probe, launched in 1994, examined areas of permanent darkness where ice might well be preserved; and NASA's Lunar Prospector (1998) may have confirmed this (still controversial) hypothesis.

PHOBOS AND DEIMOS: MARS' TINY MOONS

Leaving the Moon, we mention in passing the two small satellites of Mars: Phobos and Deimos, discovered by Asaph Hall in 1877. Apart from their heavily

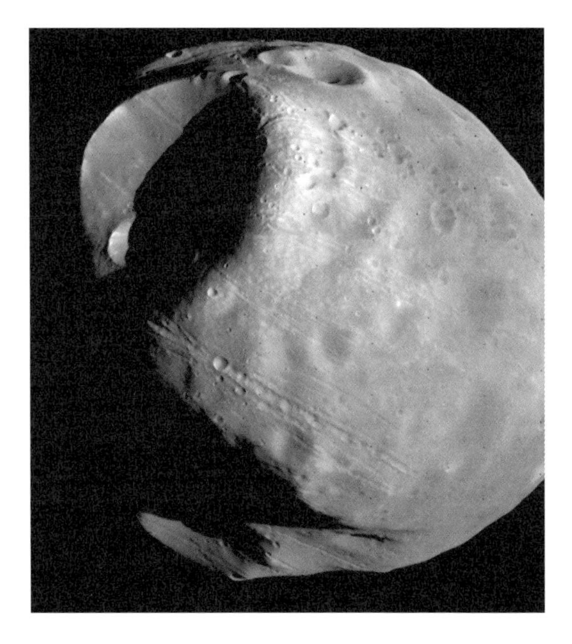

Figure 7.3. Phobos, as seen from Viking 1.

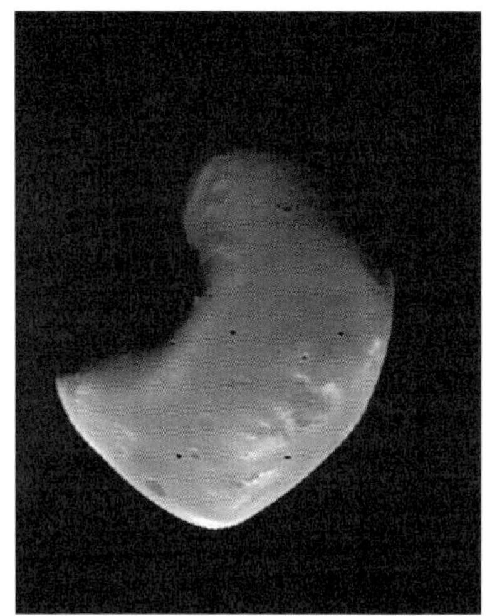

Figure 7.4. Deimos, as seen from Viking 2 at a distance of 1,400 km. Phobos and Deimos may be D-type asteroids from the outer main belt, captured by Mars' gravitational field.

cratered surfaces, they have very little in common with Earth's Moon. Phobos and Deimos are not spherical, and their longest dimensions are, respectively, 29 km and 15 km. Images of both bodies were obtained by Mariner 9 and Viking, while Phobos 2 imaged Phobos only.

The two satellites have a very low albedo, of the order of 0.05. Phobos' density is 1.9 g/cm^3, while that of Deimos is 1.7 g/cm^3. Spectroscopic measurements have shown that the two objects resemble D-type asteroids of the outer main belt. If this is their origin, than they could contain water - but this remains to be verified.

VENUS, EARTH AND MARS: THREE VERY DIFFERENT WORLDS

Let us turn now to the Earth and its two neighbours, Venus and Mars. All three are rocky bodies possessing stable atmospheres. Venus orbits the Sun at 0.7 AU, and in both density and size it resembles Earth. Mars, at 1.5 AU from the Sun, is about ten times less massive, and its density (3.95 g/cm^3) is appreciably lower. However, they all differ in their surface temperature and pressure values. Venus has a surface temperature of 730 K and a pressure of 90 bars, while Mars' average temperature is 230 K (with strong local and seasonal variations), with a surface pressure of around 6 millibars. Between these two extremes lies the Earth, where the average surface temperature and pressure of 288 K and 1 bar create an extremely fortunate environment where water can exist in all three states, as ice,

Figure 7.5. A Hubble Space Telescope image of Mars, taken in 1999. Mars' mean surface temperature is 230 K, and the mean pressure is 6 millibars. The two polar caps are visible, as well as white patches doubtless due to water ice clouds.

liquid and vapour. We shall attempt, throughout the following pages, to reach an understanding of why these three worlds, born under very similar conditions, have taken such divergent paths towards the very different environments we see today.

Let us first of all examine the atmospheric composition of the three planets. Our knowledge of our neighbours' atmospheres has come from the findings of Earth-based spectroscopy, and from spaceprobes, with *in situ* measurements carried out principally by mass spectrometers. Ignoring for the moment the vast difference in surface pressures, the atmospheres of Venus and Mars are remarkably similar. Both consist mainly of carbon dioxide (95%), the rest being molecular nitrogen with traces of argon, carbon monoxide and water vapour.

Earth's atmosphere consists mainly of N_2 (77%) and O_2 (21%), and water is present in large quantities. This is due mostly to evaporation from the oceans. If they evaporated completely their gaseous equivalent would create a surface pressure of several hundred bars! Carbon dioxide (only 0.03% by volume) was present in quantity in Earth's early atmosphere, but has been dissolved into the oceans to form carbonates. Bearing this in mind, we can see that the CO_2:N_2 ratio at the beginning of our planet's history was very similar to those of its neighbours. Earth's abundant oxygen – almost non-existent on Venus and Mars – can be explained by reference to the emergence of life. The principal (and striking) difference, however, in the atmospheric composition of the terrestrial planets is that of water content: so abundant on Earth, water is present in only minute traces (0.1%) in the atmospheres of Venus and Mars.

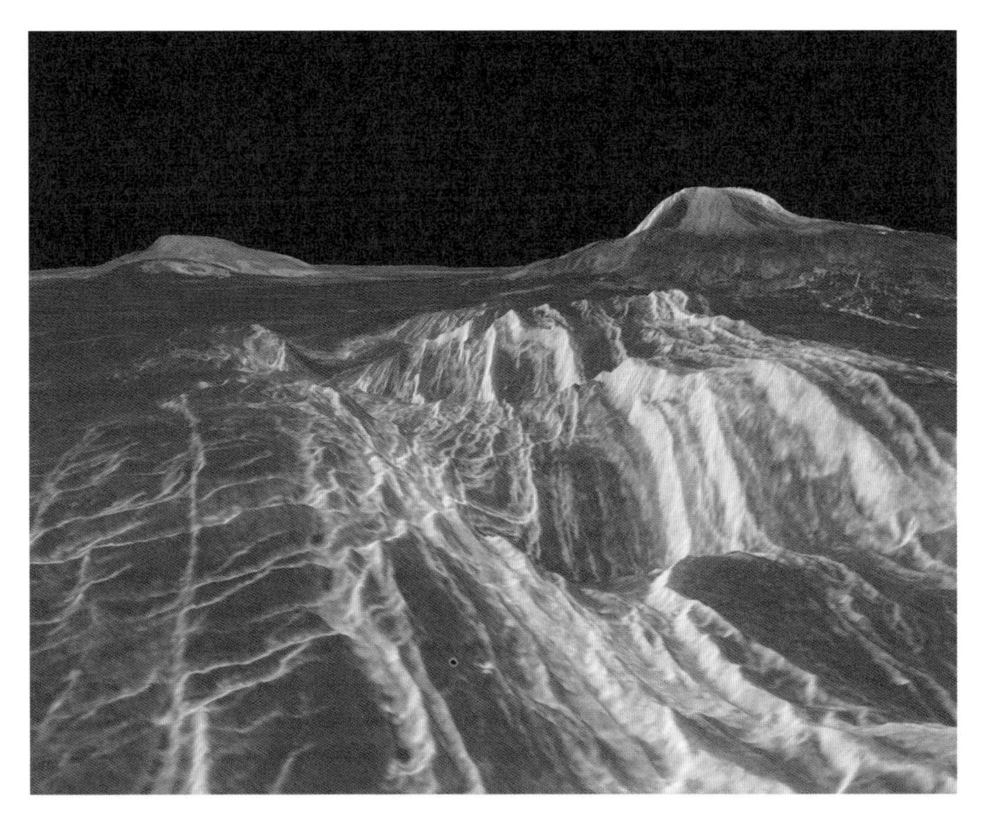

Figure 7.6. A computer-generated three-dimensional view of the surface of Venus, using Magellan data. The colours are based on images from the Russian Venera 13 and Venera 14 probes. The surface temperature of Venus is 730 K, and the surface pressure is 90 bars. (Image created by E. de Jong, J. Hall and M. McAuley, at the JPL Multimission Image Processing Laboratory.)

TRACES OF WATER VAPOUR ON MARS AND VENUS

Mars

In Chapter 2 the search for water on Mars was mentioned. In the early years of the Space Age and into the 1960s, astronomers realised that water was present in the gaseous state on Mars only in negligible quantities. The polar caps - much more likely sites - were identified as containing frozen carbon dioxide rather than the expected water ice. Data received from 1976 to 1978, from the two Viking missions, marked a great step forward. Launched respectively on 20 August and 9 September 1975, they reached Mars in 1976 and each sent down a Viking Lander module. Originally designed for a working life of 90 days, the two landers carried on functioning for years, Viking Lander 2 closing down in August 1980 and Viking Lander 1 in November 1982. The two orbiters remained active

Figure 7.7. The Earth's atmosphere, dominated by N_2 (77%) and O_2 (21%), is very rich in water, due mostly to evaporation from the oceans. Ice-crystal cloud formations are ever-present.

until 1980, and were thus able to study seasonal climatic fluctuations on Mars for two martian years. During the first of these, observations were hampered by a vast global dust storm, which veiled the surface of Mars for several months. Nevertheless, Viking data still offer a mine of information, used to this day. Until the arrival of Mars Global Surveyor two decades later in 1997, the findings of the Viking missions, fruit of a spectacular technological achievement, remained unparalleled.

An instrument carried by the Viking Orbiters, the Mars Atmospheric Water Detector (MAWD), was designed to take continuous readings of the amount of water vapour present between the orbiter and the surface. Its spectrometer measured the 1.38-μm transition of water in the near-infrared, and both orbiters confirmed that Mars' atmosphere has an unimportant water component (with a maximum of less than 0.1%). A chart was drawn up of the abundance of water as

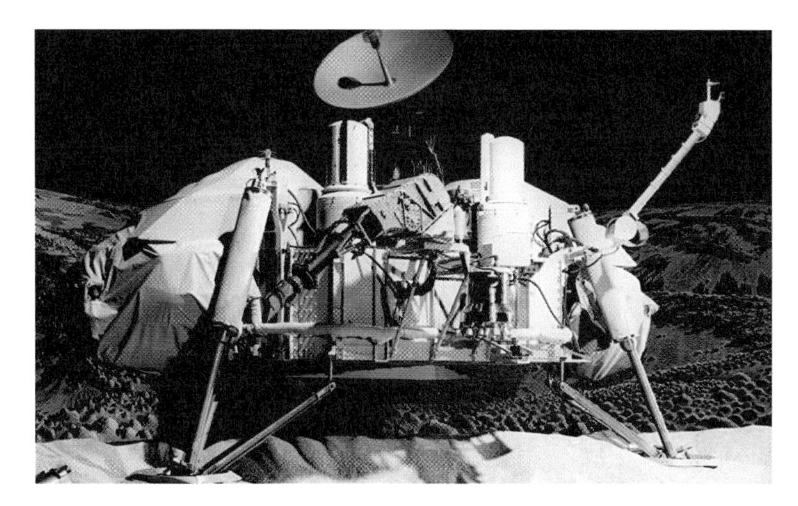

Figure 7.8. A model of the Viking Lander. These modules – carried on Vikings 1 and 2 – landed on Mars in 1976.

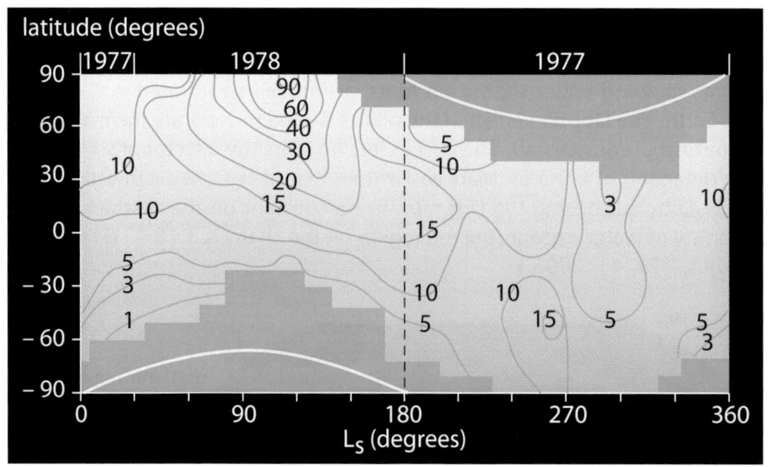

Figure 7.9. A chart showing the abundance of water vapour on Mars, averaged in longitude, as a function of latitude and areocentric longitude L_S. This parameter indicates the position of the planet in its orbit and thus defines the seasonal cycle, L_S being 0° and 180° at the equinoxes. It will be seen that the quantity of water vapour shows considerable fluctuations, with a very marked maximum at high northern latitudes when L_S = 90° to 120°, during northern spring, when water sublimates as the north polar cap melts. The same phenomenon, though less marked, is seen around the south pole when L_S = 260° to 270°. (From B.M. Jakosky and C. B. Farmer, in *Mars*, University of Arizona Press, 1992.)

a function of both latitude and seasons (Figure 7.9). The chart shows, in the northern hemisphere, a marked maximum at the pole in spring and early summer; and in the southern hemisphere, a similar but more modest maximum half a season later. The explanation for this phenomenon lies in the sublimation of a certain quantity of water ice, trapped at the poles beneath the frozen carbon dioxide during the martian winter (Figure 7.9a).

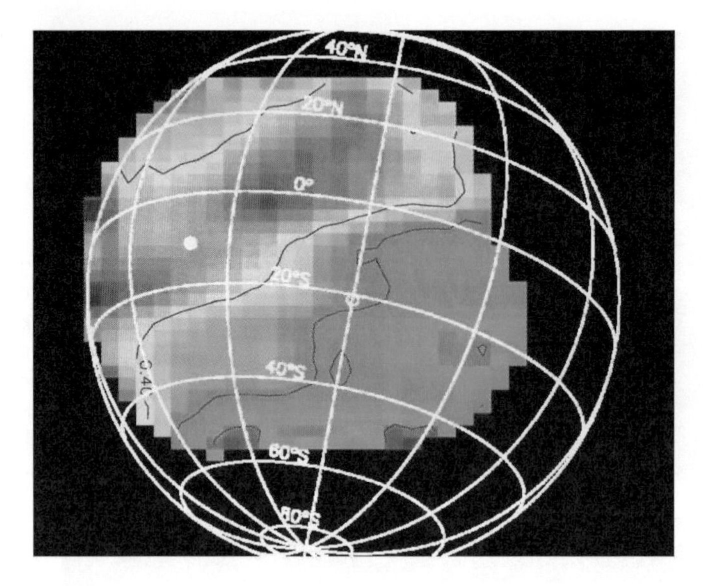

Figure 7.10. A chart showing water vapour on Mars, as imaged by infrared spectroscopy from Earth. The Doppler effect of the planet is used to separate the martian HDO line from its terrestrial counterpart. The D/H excess on Mars (by a factor of six – see p. 153) makes the detection of HDO on Mars easier than that of H_2O. We can estimate the abundance of H_2O by supposing the D:H ratio to be constant on the martian disc. The maximum quantity of water vapour (the red region on the chart) is 3.10^{-4} . (From T. Encrenaz *et al.*, *Icarus, 179*, 43, 2005.)

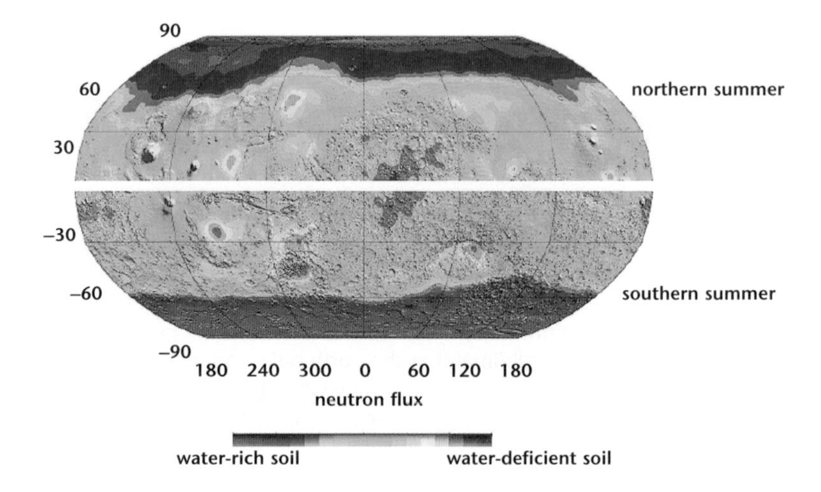

Figure 7.11. A chart showing neutron flux indicating the hydrogen-atom content of the martian surface to a depth of 1 metre, based on measurements with the neutron detector on board Mars Odyssey. The blue and violet areas indicate high ice content; and the green, yellow and red areas, low ice content. The presence of hydrogen can be explained only by water ice being trapped below the surface.

There are other indications of the presence of water on or beneath the soil of Mars. For example, infrared spectra of the martian surface, taken regularly through telescopes over several decades, have shown the signature of intense hydration at 3.0 and 3.5 μm. This signature is characteristic of the existence of OH in the soil. More recently, Mars Odyssey used another technique (a neutron spectrometer detecting high-energy particles) to reveal the presence of great quantities of hydrogen atoms (and by inference, water ice) near the poles, less than 1 metre below the surface (Figure 7.11). Two years later, Mars Express turned its infrared spectrometer towards the south pole and confirmed the result, detecting water ice beneath the permanent CO_2 cap. So, as will be discussed below, there are various indicators of the existence of sub-surface water as permafrost, suggesting that liquid water might once have flowed early in the history of Mars.

Table 7.1. Comparative table of planetary atmospheric composition

Planet	Surface pressure (bar)	Surface temperature (K)	Main atmospheric constituents	
Venus	90	730	CO_2	(0.965)
			N_2	(0.035)
			CO	(0.001 at z = 100 km)
			H_2O	(50 ppm at z < 50 km)
			SO_2	(150 ppm at z = 20 km)
Earth	1	288	N_2	(0.77)
			O_2	(0.21)
			H_2	(0.017)
			Ar	(0.01)
			CO_2	(0.0035)
			H_2O	(0–004)
			CO	(10 ppm at z = 90 km)
			O_3	(7 ppm at z = 30–40 km)
			CH_4	(1 ppm at z < 30 km)
			NO_2	(0.6 ppm at z = 0 km)
Mars	0.006	218	CO_2	(0.95)
			N_2	(0.027)
			Ar	(0.016)
			O_2	(0.0013)
			CO	(0.0007)
			H_2O	(0.0003, variable)

ppm = parts per million

Venus

Searching for water on Venus is a radically different exercise. Unlike Mars, Venus is cloaked in a dense layer of clouds, which conceal its surface completely. These clouds, at an altitude of about 50 km, are composed of sulphuric acid. Moreover, there is acid rain, falling towards a surface subject to crushing pressure and baked in the torrid heat. It would be difficult to imagine a more uninviting world! What is true for humans should also be true for robot probes: dropping a Venus lander onto the surface from an orbiting spacecraft might seem too much of a challenge, but it was taken up by the Soviets. In the 1970s and 1980s, the Intercosmos programme sent a whole series of probes to Venus. The first to descend successfully to a soft landing was Venera 7, in 1970, and the last of the series to land was Venera 14, in 1982. The two modules of the American Pioneer Venus mission, launched in 1978, also studied the venusian surface. Other orbital probes continued the work – notably Venera 15 and Venera 16, and Magellan, which carried out radar mapping of the surface.

To these *in situ* explorations must be added Earth-based spectroscopic measurements in the near-infrared, from 1990 onwards, and data from the Galileo probe, which flew past the planet in February 1990. How was the nature of the deep atmosphere of Venus, beneath the cloud layer, revealed? The method relied on observations of the night side, to avoid problems with sunlight reflected from the clouds. Observations of carefully chosen spectral windows, between the absorption bands of carbon dioxide, showed very weak thermal radiation originating in the lower atmosphere and from the surface. These windows, located mainly at 1.7 and 2.3 μm, allowed the examination of the chemical composition of the lower atmosphere, and, in particular, indicated the abundance of H_2O and its isotope HDO (Figure 7.12). The European Venus Express mission, launched in 2005, will produce a more accurate map of Venus' atmospheric constituents. Measurements showed that water vapour is present in quantities of 30 parts per million. Given the temperatures on the planet, there is no other possible reservoir of water, and the atmosphere of Venus is therefore extraordinarily dry.

THE HISTORY OF WATER VAPOUR ON MARS AND VENUS

Although Venus is devoid of water today, it was not always so, as measurements of the abundance of heavy water HDO have proven. The first readings from probes on the surface produced a spectacular result: the D/H ratio, deduced from the ratio HDO:H_2O, seemed to be 100 times greater than on Earth! The standard Earth value (SMOW – Standard Mean Ocean Water), based on unambiguous results from sea water, is $1.5/10^{-4}$. In 1990, infrared spectroscopic observations (see above) produced a more accurate value: the deuterium ratio on Venus is 120 times greater.

How can this deuterium excess on Venus be explained? The usual answer nowadays is that there has been considerable outgassing of water vapour, which was photodissociated in the upper atmosphere. Hydrogen was light enough to escape from the planet's gravitational field. The deuterium atoms, twice as heavy, could not escape as easily – hence the deuterium excess. This scenario suggests that during its early history Venus must have had vast amounts of water, perhaps in quantities comparable to those on Earth.

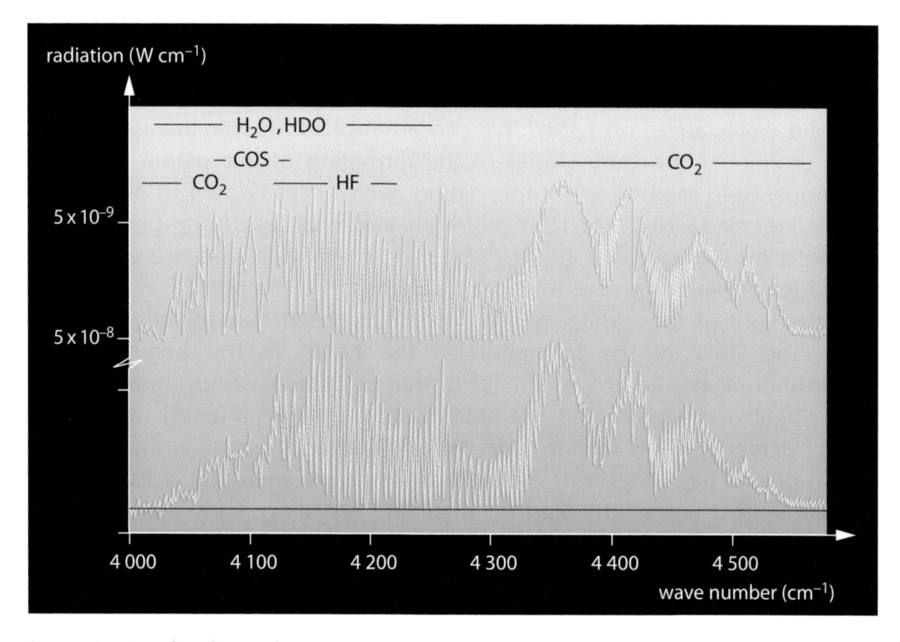

Figure 7.12. The thermal spectrum of the night side of Venus, observed in the near-infrared by the high-resolution spectroscope of the Canada–France–Hawaii telescope on Mauna Kea. The upper spectrum is based on calculation, and the lower spectrum is that observed. Between the strong absorption bands of CO_2 we see the presence of absorption bands due to CO, H_2O, HDO, OCS and SO_2. Due to the simultaneous observation of H_2O and HDO, the D:H ratio can be measured. The excess is found to be 120. (From B. Bézard *et al.*, *Nature, 345*, 508, 1990.)

The same story can be told of Mars. In 1989 the D:H ratio was measured for the first time, by infrared spectroscopy from Earth. Here too, a deuterium excess (this time by a factor of 6) was detected (Figure 7.13). It is not surprising that Mars has a smaller deuterium excess than Venus, because the martian atmosphere is far less massive. As was the case on Venus, this excess points to substantial outgassing into Mars' primitive atmosphere. Its surface pressure may once have exceeded 1 bar – still a much smaller value than that of the early atmospheres of Venus and the Earth.

DIVERGENT DESTINIES

Venus, Earth and Mars formed under very similar conditions, but have since evolved in very different ways. Several reasons can be invoked, but the most compelling is the presence of the water molecule. Why is this? The physical conditions present in the primitive atmospheres of the three planets were such that, on Venus, water was in the gaseous state; on Earth, it was liquid; and on Mars, solid. A small change in temperature, altering one molecule, set off a vast

What is the origin of the water in the Earth's oceans?

How did water arrive on Earth? We can attempt to answer this question by again examining the early stages of the formation of our planet. The most commonly used method is isotopic dating, which uses two types of indicator: non-radiogenic isotopes of relatively light elements, and long-period heavy elements; for example, the (^{87}Rb, ^{87}Sr) pair. These indicators are used in the case of terrestrial, lunar and meteoritic samples (see Chapter 3).

In the light of these data, geochemists have been able to tell, in general terms, the story of the formation of the Earth. In the beginning, the composition of the Earth was similar to that of carbonaceous chondrites, the most primitive meteorites in the solar system (see Chapter 6), and rich in volatile elements. A few million years on, and the youthful Sun embarked on its T Tauri phase (see Chapter 3). The intense and very energetic radiation had a reducing effect upon the Earth's material, expelling light elements. The mechanism of planetary formation proceeded by accretion of planetesimals, and after a few tens of millions of years the Earth had acquired about 85% of its total mass. Then came the giant impact, when the young Earth collided with a smaller planet the size of Mars, resulting in the formation of the Earth–Moon system. It was this collision that set off the internal differentiation of the Earth, leading to a nickel–iron core and the separation of the mantle into two layers, each with its own convective movements. An effect of this separation is the retention of volatile elements within the lower mantle. In the case of Venus, the interior undoubtedly possesses a single convective system.

But the early Earth was still without water. Where did the water in its atmosphere and its oceans come from? Firstly, it was a product of outgassing from the upper mantle. The composition of the gas was similar to that of protosolar abundances, and sampling hot spots on the volcanoes of Hawaii still provides evidence of this. Secondly, water came from meteoritic impacts, which were especially frequent during the first 100–200 million years of Earth's history. The composition of the meteorites was certainly close to that of carbonaceous chondrites, to which we must add a cometary component. The D/H value in Earth's oceans (1.5×10^{-4} – six times the protosolar value) can therefore be explained via a combination of an outgassed component (of value close to the protosolar), a (certainly predominant) component of carbonaceous chondrite type (D/H = 1.5×10^{-4}), and a cometary component (D/H = 3×10^{-4}) (see Chapters 4 and 6).

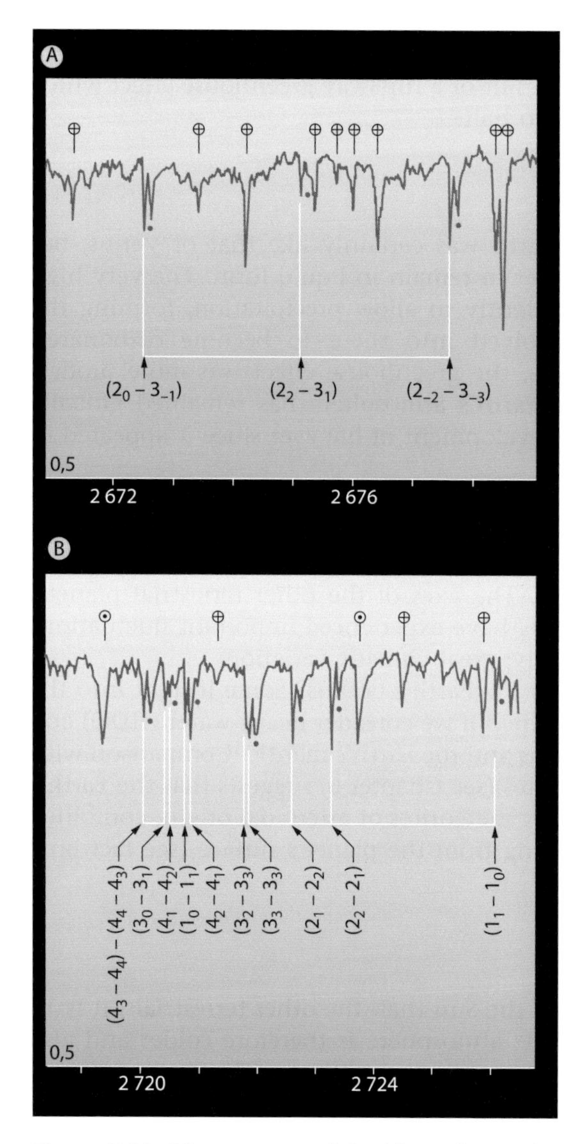

Figure 7.13. The spectrum of the Mars, observed at high resolution by the Canada–France–Hawaii telescope on Mauna Kea. Several lines due to heavy water, HDO, are present. By comparing this with the estimated quantity of H_2O present on the planet at the same epoch, the first determination of the D:H ratio was made possible. The excess, compared with the value for Earth, was a factor of 6. Lines due to the Earth's atmosphere are shown with the symbol for Earth. (From T. Owen *et al.*, *Science, 240*, 1767, 1988.)

and unstoppable process. This should give us pause for thought about the consequences that a rise in temperature might have for our own planet.

Venus

Orbiting the Sun at a distance of 0.7 AU, Venus has been at temperatures above 300 K since it was first formed. Its water has therefore been in the form of vapour. As we have seen, it was once abundant – perhaps as abundant as the planet's carbon dioxide. Water vapour and CO_2 are both very efficient greenhouse gases: their intense vibrational bands, in the infrared, bear witness to their ability to absorb solar radiation and convert it into infrared in the form of thermal energy (see Introduction). The solar flux – converted into thermal energy and trapped between the surface and the lower atmosphere – cannot escape to space, and the greenhouse effect increases. The surface temperature then rises, climbing to the current value of 730 K. But what has become of the atmospheric water? It has almost certainly been photodissociated. The hydrogen has escaped, as suggested by the deuterium excess. The heavier oxygen could not have escaped in the same way, and its disappearance remains enigmatic. It may have associated itself with other constituents such as chlorine, but there is no evidence of this. In summary, Venus is a prime example of an extraordinarily hostile planet – the result of a runaway greenhouse effect which no other mechanism has been able to halt.

Earth

The primitive atmosphere of the Earth was certainly like that of Venus, but physical conditions allowed the water to remain in liquid form. The very high temperature at the outset fell sufficiently to allow precipitation, forming the oceans. Carbon dioxide then dissolved into these to become carbonates. Deprived of its two principal agents, the greenhouse effect was more modest on Earth. The temperature of the Earth's atmosphere has remained globally stable, and this has promoted the development of life ever since it appeared in the ocean depths.

There is another factor, peculiar to the Earth, which has contributed to the constancy of its temperature: the presence of the Moon. Calculations based on celestial mechanics reveal that the Moon has played a very important part in the stabilisation of the Earth's axial tilt. The axes of the other terrestrial planets, which lack any such massive satellite, have experienced important fluctuations in their orientation and, consequently, great climatic variations.

What is the origin of the water in the Earth's oceans? Some insight into the answer to this question may be obtained if we consider heavy water (HDO) and the D:H ratio (yet again) in the oceans and the Earth's mantle. Comparison with this ratio in various kinds of meteorites (see Chapter 6) suggests that the Earth's water derives mainly from impacts by meteorites of asteroidal origin, combined with cometary impacts and outgassing from the planet's surface (see fact box, p. 154).

Mars

Now we come to Mars. Further from the Sun than the other terrestrials, it is 10 times less massive than the Earth. Its atmosphere is therefore colder and also more tenuous than the atmospheres of the Earth and Venus. The atmospheres of the terrestrial planets originated at least partly from volcanic activity, with a contribution from meteoritic impacts: the larger the planet, the greater will be the contribution of outgassing, and the more radioactive elements there will be to provide the energy to drive the outgassing process. Because Mars' gravitational field is weaker, it has also attracted fewer meteorites. Given these facts, it is hardly surprising that the primitive atmosphere of Mars was so much less dense than those of its two nearest neighbours. Mars' internal energy was greater during its early history. Evidence of this is the residual magnetic field associated with its most ancient terrains, and discovered by the magnetometer of Mars Global Surveyor. Mars once possessed its own internal 'dynamo'. This internal activity, responsible for the martian volcanoes, continued for more than 1 billion years, then progressively died away. Due to the falling temperature the greenhouse effect faded, and the water in the martian atmosphere ended up trapped within the surface as ice or permafrost (of which more later). What happened to the carbon dioxide in the primitive atmosphere of Mars? This question remains unanswered. It may have been taken into the surface in the form of carbonates, as was the case on Earth, but, despite repeated studies, this has never been satisfactorily proved.

THE HISTORY OF WATER ON MARS

There are various indications to suggest that water was present on the surface of Mars early in its history. Firstly, the deuterium excess points to a primitive atmosphere much denser than today's, implying the possible liquefaction of water. Then there is the existence for a billion years of the 'dynamo' – a sign of the internal activity responsible for a denser and warmer atmosphere in the past.

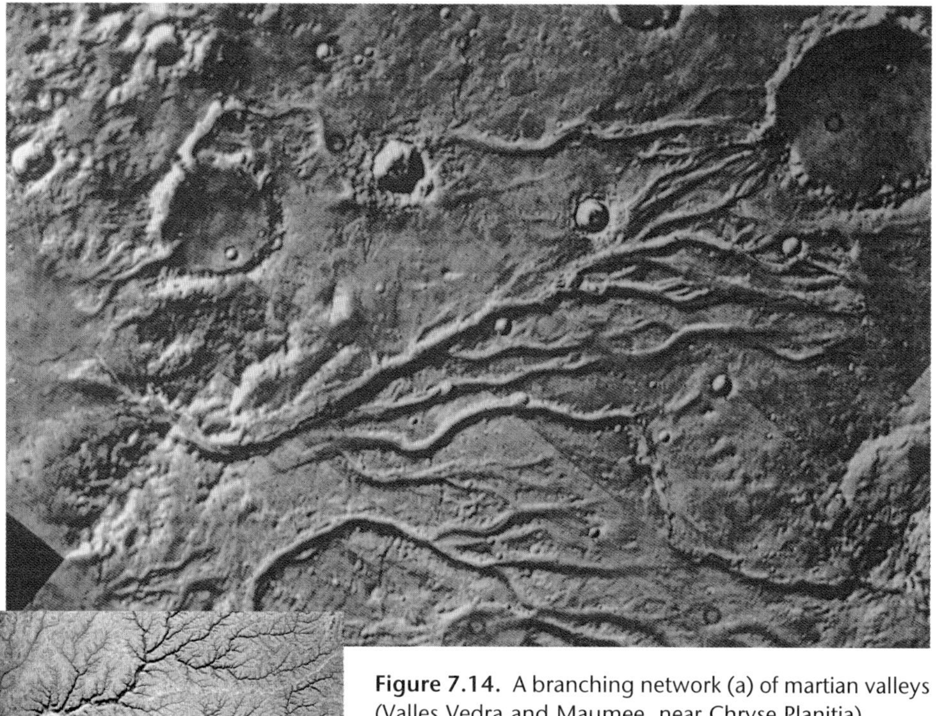

Figure 7.14. A branching network (a) of martian valleys (Valles Vedra and Maumee, near Chryse Planitia) compared with the Hadramaut Plateau (b) in South Yemen, photographed from the Space Shuttle *Discovery*. Both images show an area about 100 km wide. The similarity of the two structures is striking, though the scales are different. The martian valleys are noticeably wider. (NASA and 41G-36-36; courtesy of Earth Sciences and Image Analysis Laboratory, NASA JSC.)

Figure 7.15. An outflow valley in Ares Vallis. Here, 100,000 m^3 of water per second are thought to have flowed in an outpouring of 10^{13} m^3 of underground water (from permafrost) under pressure. This image was taken by the Themis camera onboard Mars Odyssey in 2002. Illumination is from the left.

Again, we have seen that certain surface minerals have been hydrated, and there is water ice beneath the poles. The morphology of the martian surface, as revealed by the cameras of the Viking probes and Mars Global Surveyor, provides another set of clues. The most spectacular of these is the discovery of many branching valleys in areas of ancient land-forms, and outflow channels in chaotic terrain, which mark the passage of flowing water (Figures 7.14 and 7.15.). Also, the existence of lobed ejecta around some impact craters (Figure 7.17) seems to indicate the presence of viscous (or even liquid) sub-surface material, splashed outwards at the time of impact. The water which flowed across Mars may well be of sub-surface origin, from hydrothermal sources or underground water tables (Figure 7.16).

Another piece of information comes from the Mars Orbiter Laser Altimeter (MOLA) on board Mars Global Surveyor. This instrument carried out a very accurate topographical survey of the whole martian surface. We already knew from Viking and Mariner 9 that this surface may be divided into two distinct parts: to the south, highland areas with extensive cratering, and therefore very ancient; to the north, low volcanic plains with, at the equator, the Tharsis Plateau. The MOLA determined the precise altitude of the boundary (line of dichotomy) between these two regions, and showed that it is constant over distances of the order of 1,000 km (Figure 7.18). Are we seeing here an ancient ocean shoreline?

We can try to reconstruct the history of a possible primaeval ocean. It is probable that the water of the terrestrial planets, and the rest of their atmospheres, originated partly from outgassing and partly from impact events. The latter were particularly intense during the first billion years of the history of the solar system. At the end of this period of bombardment, 3.8 billion years ago, the sub-surface material of Mars must have been particularly rich in water. Mars' internal energy, responsible for volcanism and hydrothermal activity, was able to maintain the circulation of liquid water on the surface for a certain time. This was the period when the branching valleys of the southern hemisphere were formed.

Has liquid water flowed on Mars for long periods during a more recent past? This is currently a subject of debate. Altimetry from Mars Global Surveyor, showing a possible shoreline between the higher plateaux in the south and the northern plains, has led some scientists to suggest that an ocean once covered these plains. However, the absence in these regions of hydrated material, as revealed by infrared measurements taken by Mars Express, seems to invalidate this hypothesis. Water has certainly flowed on Mars in the recent past, as the outflow channels near Valles Marineris testify. The presence of liquid water also lies at the origin of sulphates, observed simultaneously in the Meridiani Planum by the Mars rover Opportunity, and in the Marineris region by infrared spectroscopy from the Mars Express orbiter. These flows could also have been the result of relatively recent and localised (and undoubtedly volcanic) phenomena. They do not necessarily imply the prolonged presence of liquid water on Mars in the recent past. This question – which is still the subject of very lively debate – is crucial in our understanding of exobiology: the long-term presence of liquid water is an *a priori* factor favouring the emergence of life. These intriguing topics will continue to lie at the heart of our future exploration of Mars.

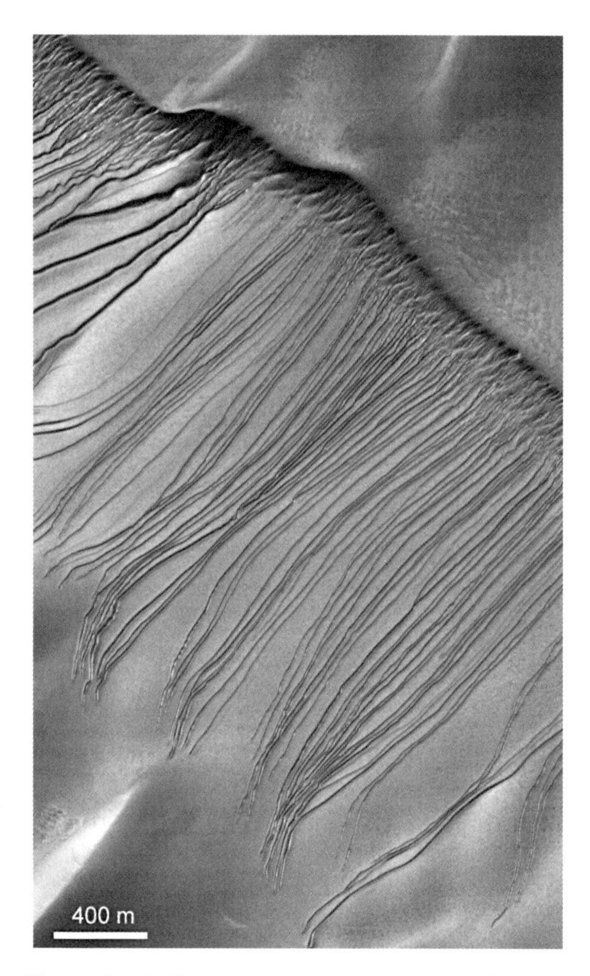

Figure 7.16. Flow traces on martian dunes. These features tend to contradict the theory of sub-surface sources.

SEARCHING FOR LIFE ON MARS

The search for life on Mars goes back a long way. The first observations from space may have dispelled once and for all the myth of the canals, but the topic is

Figure 7.17. A night-time image, taken by the Themis spectrometer onboard Mars Odyssey in December 2002, of lobed ejecta impact craters in the Isidis Planitia region. In the plains of Mars' northern hemisphere, the ejecta from this type of crater are particularly far-flung, suggesting the presence of great quantities of water ice at the time of impact.

still a lively one. NASA embarked upon its Viking mission with a view to searching for signs of life on Mars. Viking's scientific findings were many, but the avowed primary objective drew a blank. What was this objective? Three experiments mounted on the landers were designed to search for evidence of biological activity, while a fourth looked for organic compounds. This last experiment produced completely negative results, while the biological investigations were inconclusive. The consensus was that biological signatures were absent, and responses obtained by the bio-experiments were attributed to the presence in the soil of a powerful oxidant, which could also explain the lack of organic molecules. The oxidant proposed is hydrogen peroxide (H_2O_2), which is not very abundant on Mars, but was detected with Earth-based infrared and submillimetre spectroscopy in 2003.

For two decades the debate slumbered; then in 1996 it was much enlivened by a NASA announcement. Sensation! A meteorite from Mars had been discovered, with possible traces of fossil life within it. What kind of meteorite was this? More than twenty meteorites thought to be of martian origin have been found. They are known as SNC meteorites, after the groups to which they belong: Shergottites, Nahklites and Chassignites. This identification is based upon the similarity of their isotope ratios to those measured *in situ* on Mars by the Viking probes. The meteorite in question is one of these, and is designated ALH 84001. It was ejected from Mars about 16 million years ago, and fell in Antarctica 13 million years ago. At 4 billion years old, it predates the other martian meteorites by a long way. Within its basaltic texture are globules of magnesium carbonate which may have formed in water 3.6 billion years ago. Inspection of these globules has revealed structures resembling nanobacteria, and claims have been made for the presence of polycyclic aromatic hydrocarbons (PAHs) and tiny grains of oxides and iron sulphate analogous to those produced by Earthly bacteria. Some researchers have seen these features as signs of fossil life-forms in this meteorite. After several years of

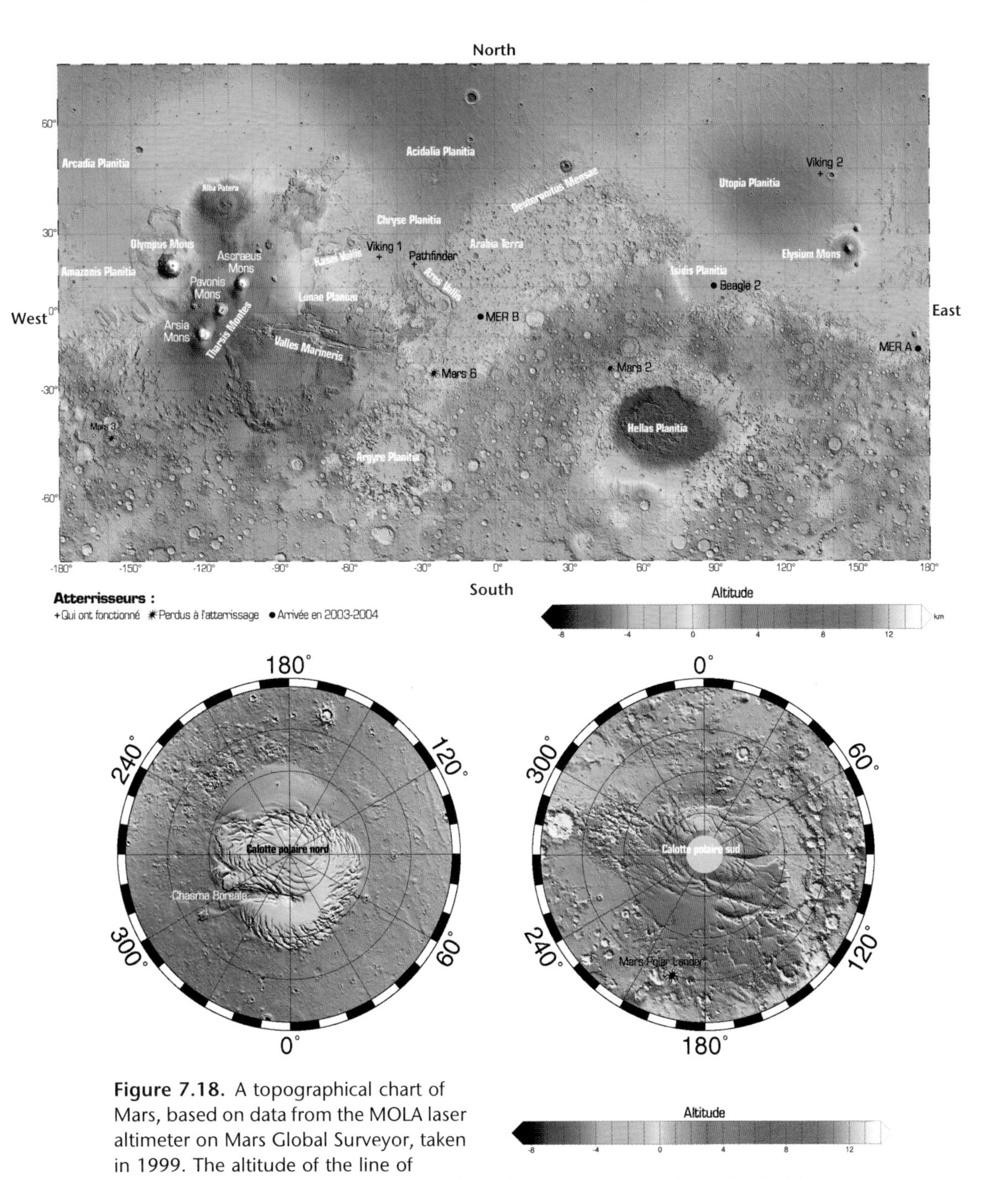

Figure 7.18. A topographical chart of Mars, based on data from the MOLA laser altimeter on Mars Global Surveyor, taken in 1999. The altitude of the line of dichotomy, separating the northern and southern regions, was determined with great accuracy. It is remarkably constant over distances of the order of 1,000 km. It can be seen that the more heavily cratered southern hemisphere is somewhat more elevated than the plains of the north. (From F. Forget *et al.*, *La Planète Mars*, Belin, 2003.)

controversy, it now seems that there may be other explanations not involving the hypothesis of ancient life on Mars.

What direction will the search follow in the future? NASA has drawn up a programme of exploration, first of all targeting those sites where water may well have flowed, and where traces of fossil life might be found. European scientists, on the other hand, are focusing primarily on orbital surveys, to be followed by *in situ* experiments using a network of several landers at different sites on the martian surface. This approach will allow for a seismographic analysis of the planet's interior and eventually the dispatch of martian material to Earth. The analysis of lunar meteorites and rock samples has amply demonstrated how a wealth of information can be extracted from such objects under laboratory conditions here on Earth, using the most sophisticated instruments. Will people go to Mars within a few decades? It is not certain that science will profit from this, since the potential of robotic exploration is far from being exhausted. The cost of a manned mission is beyond all comparison with that of a robotic probe; yet the same was true thirty-five years ago, when men went to the Moon. The arguments for such an adventure revolve, of course, not merely around science. If the mission cannot be justified from the scientific point of view alone, it is certain that science will still extract the maximum benefit from it.

The future exploration of Mars

The year 2003 was particularly auspicious for the exploration of Mars. Three space missions were launched in June and July – one by ESA (Mars Express), and two by NASA, carrying Mars roving vehicles. The Mars Express mission consisted of an orbiter carrying seven experiments, and a lander – Beagle 2 – produced in the United Kingdom. Onboard remote-sensing equipment included a camera, infrared spectrometers, solar and stellar occultation experiments, an energetic neutral atoms analyser, a radar instrument, and a radio science experiment. Mars Express successfully entered Mars orbit on 24 December 2003. The Beagle 2 lander, scheduled for an 180-day mission on the surface, unfortunately failed to initiate. The NASA rovers, Spirit and Opportunity, landed safely in January 2004. They carry experiments designed to seek out water (present or past) at the two selected sites – Gusev Crater for Spirit, and the Meridiani Planum region for Opportunity. Among their instruments is an infrared spectrometer working between 5 and 29 μm, specifically to investigate any signs of water. The American landers send their signals to the orbiting craft and to Mars Express, from where they are relayed to Earth.

Another probe, Nozomi, was also *en route* for Mars at this time. Launched by the Japanese Space Agency in July 1998, its mission was abandoned as a result of instrumental problems. It carried experiments aimed at examining the interaction between the martian atmosphere and the solar wind.

At the time of writing (January 2006), Europe's plans for further Mars missions are still uncertain. NASA maintains a rolling programme of launches every two years – the frequency being determined by the available launch windows for Earth–Mars flights, which in turn depend on the configuration of the two planets in their orbits. A general objective is the search for present or past water, and there is always the hope that fossil traces of life might be found. On 12 August 2005, Mars Reconnaissance Orbiter (MRO) was launched. This spacecraft carries an atmospheric probe, a very-high-resolution camera, and a radar device. Another mission aimed at exploring the surface will leave in 2007. Lastly, the Mars Science Laboratory (MSL) is scheduled for departure in 2009. It is 3 metres long and weighs 1 tonne, and is therefore much larger than previous Mars probes. It will carry a whole range of instruments designed to study traces of the liquid water once present on Mars.

Figure 7.19. An artist's impression of Mars Express in orbit.

8
The search for other Earths

Are we alone in the Universe? The exploration of the different objects within the solar system suggests that the probability of finding extraterrestrial life in the vicinity of our star is very small indeed. Since 1995, the discovery of more than 150 extrasolar planets (exoplanets), beyond the confines of the solar system, has given a boost to the search for other life-forms. Could life have appeared on any of those planets? How might we try to discover it? The search for liquid water will be a vital criterion.

The quest for extraterrestrial life is not of recent origin. The question 'Are we alone in the Universe?' is probably as old as humanity, and is evident in scientific and literary debates throughout the centuries. During the last decade our perception of extraterrestrial life, or exobiology, has taken on a new dimension – due to the discovery of more than 150 extrasolar planets.

The search for life elsewhere in the Universe raises many questions. First, the very nature of life itself. We know of only one example. What are we supposed to look for if it is outside our experience? Second, the emergence of life. We still do not know how it came to be here on Earth, but we are certain that the existence of liquid water played a major role in its development. A third question is that of the possibility of extraterrestrial life within our solar system. We have analysed various bodies and identified likely sites, with a view to the possible presence of fossil life on Mars, the existence of prebiotic chemical activity on Titan, and the possibility that there might be an ocean of liquid water beneath the surface of Europa. Leaving aside these particular cases, the most probable conclusion is that no extraterrestrial life exists in the neighbourhood of our star.

The quest for life elsewhere now has a new dimension with the recent discovery of exoplanets orbiting nearby stars. New questions have been raised. Could life have appeared on those planets? If it did, what criteria do we use to select the most likely candidates? How do we actually discover whatever kind of life might be there?

HOW DO WE DEFINE LIFE?

Before we reflect upon extraterrestrial life, it is useful to define what life is. Looking at life on our planet (the only example we have) biologists use certain criteria: *self-reproduction* (normally in identical form); *evolution through* (accidental) *mutation*; and *self-regulation* to fit the ambient environment (ensuring growth and self-preservation).

Biology teaches us that to fulfil the aforementioned functions, all living systems employ the same types of molecule: mainly nucleic acids and proteins. The chromosomes, consisting of the nucleic acids, transmit genetic information via deoxyribonucleic acid (DNA), which is built from a sequence of nucleotides

---

An artist's impression of the six telescopes of the future Darwin mission.

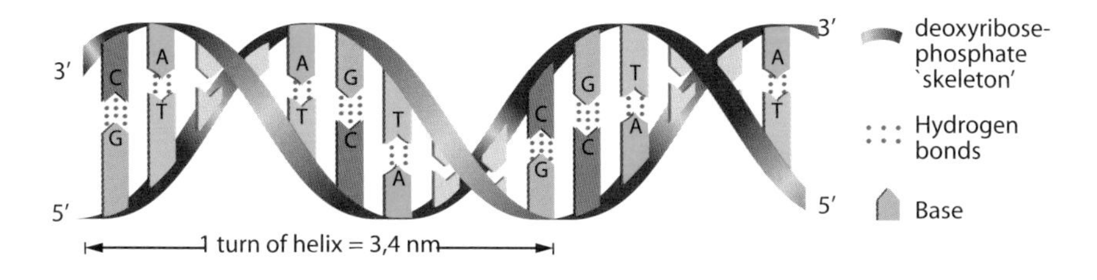

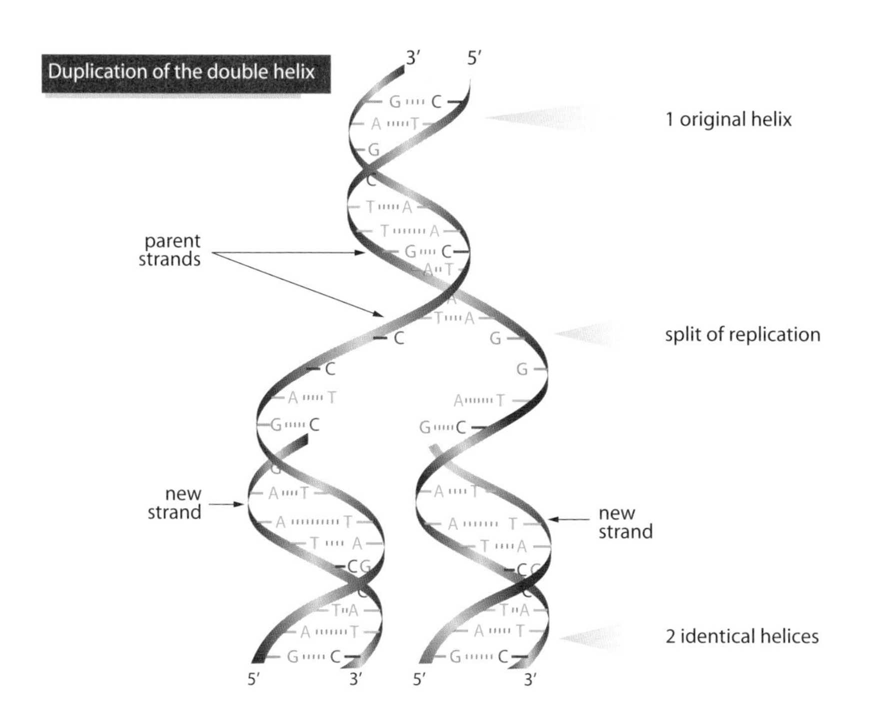

Figure 8.1. (a) Diagram of the double-helix structure of the DNA molecule. The two strands are antiparallel and complementary. (b) Diagram of the duplication of the genome. Each strand of the 'parent' double helix acts as a matrix for the synthesis of its counterpart. The 'daughter' helices consist of a parent strand and a newly formed strand.

containing only four macromolecules, or bases: cytosine, guanine, thymine and adenine. Proteins are composed of about twenty amino acids only (Figure 8.1).

HOW DOES LIFE BEGIN?

In 1953 an historic experiment took place: the first laboratory synthesis of amino acids. Following the work of the Russian scientist Aleksandr Oparin, a pioneer in the field, Americans Stanley Miller and Harold Urey succeeded for the first time

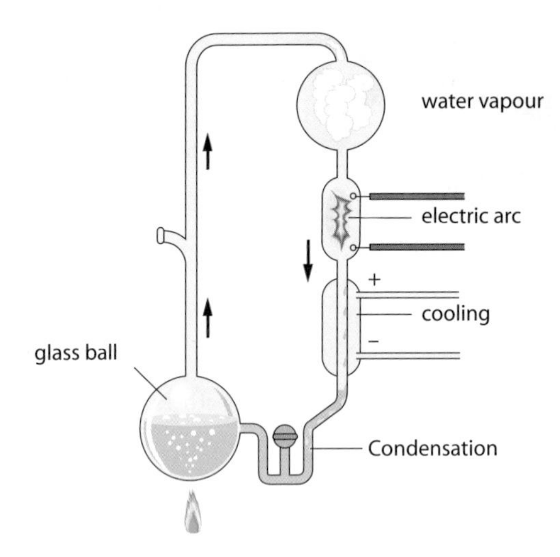

Figure 8.2. Miller–Urey apparatus, simulating prebiotic chemical evolution. This historic experiment, carried out in 1953, showed that it is possible to synthesise organic and even prebiotic molecules in a gaseous mixture of hydrogen, methane, ammonia and water vapour subjected to electrical discharges. (Modified from P. de la Cotardière, *Dictionnaire de l'Astronomie*, Larousse, 1999.)

in producing amino acids in the presence of liquid water, in a reducing atmosphere (of H_2, CH_4 and NH_3) subjected to electrical discharges. In later decades, the four DNA bases were synthesised (Figure 8.2). However, this did not answer the question of how life began. What stages were necessary between the amino acids and the living cell? Fifty years on, the answer is still not known, and the bridge between prebiotic and living matter remains to be found.

Uncertainty still surrounds the emergence of life on Earth, and we therefore cannot estimate the probability of its appearance. What is more, it seems that the primitive atmosphere of the Earth, as we have seen, was not a reducing one but contained large amounts of CO_2 and H_2O. Life could have appeared at the bottom of the oceans, around hydrothermal sources; or, according to the 'panspermia' theory, it might have been brought by infalling micrometeoric material, the debris of comets. It is impossible to put these theories to the test, since tectonic movements and erosion by water and wind have erased the traces of the emergence of life. All we know for certain is that life appeared more than 3.5 billion years ago, judging by the evidence of stromatolites - calcareous structures built up through the activity of marine organisms. To summarise: we do not know if the presence of liquid water in contact with complex molecules necessarily leads to the beginnings of life, but we do know that on Earth it made its emergence possible; and there is nothing to indicate that this phenomenon is exceptional.

Laboratory experiments show that given a source of energy (UV radiation, energetic particles, and so on), organic chemistry tends to evolve to a state of greater complexity. This is also demonstrated by the interstellar medium, in which we have detected organic molecules of more than a dozen atoms, and even more complex organic structures such as polycyclic aromatic hydrocarbons (PAHs), and the refractory residues coating interstellar grains irradiated by the ultraviolet flux from stars, or by cosmic radiation. Lastly, various amino acids, and other prebiotic compounds, have been found associated with some meteorites. It is therefore possible to envisage a carbon-based 'cosmo-chemistry' on an interstellar scale. Could life exist elsewhere in the Universe? Remember that the Sun is a very ordinary star among the tens of billions that populate our galaxy. Is ours the only inhabited world? If it is, then why?

EARLY DISCOVERIES OF EXOPLANETS

In order to have the necessary conditions (as far as we understand them) for life to begin, it is not sufficient just to find prebiotic molecules in the interstellar medium. Perhaps (or certainly?) a water environment is required, presupposing certain levels of temperature and pressure. Where might we find such conditions, if not on possible exoplanets? Until the 1990s the question was academic. Throughout the previous decade there had been many attempts to detect low-mass objects orbiting nearby stars, but they had all drawn a blank. The reason was that the techniques used at the time lacked the required sensitivity.

What were these techniques? One limiting factor stands out: a Jupiter-sized planet orbiting a few AU away from a Sun-like star cannot be detected by normal imaging methods, even if that star is comparatively near us. In the visible part of the spectrum, radiation from the planet is swamped by the light of the star (more than 10^9 times brighter). In the infrared, at around 10 μm, the contrast is less unfavourable (10^6), but still prohibitive.

Other methods are needed, based on astrometry. If a companion object is orbiting a star, the two will revolve around their common centre of gravity. The companion is too faint to be observed, but the motion of the star exhibits an oscillating movement around the centre of gravity of the system, with a period matching the time taken for the companion to orbit the star. A tiny movement indeed, but measurable with modern techniques. Two of these are astrometric measurement in relation to other stars, and measurement of radial velocity, the latter of which is also affected by the periodic movement (except in the unfavourable case where the star–planet system lies in a plane perpendicular to the line of sight) (see fact box, p. 170). Throughout recent decades the astrometric method has been preferred; but without success, as the amplitude of the motions was so small as to be undetectable by the instruments available at the time, with measurements requiring, in particular, a high degree of stability on time-scales of more than ten years. Jupiter, to take a local example, takes nearly twelve years to complete one orbit.

In the end, the radial-velocity method (velocimetry) proved the more profitable. In 1995 Michel Mayor and Didier Queloz, of the Geneva Observatory, announced a discovery which became an immediate landmark in the history of astronomy. At the Haute-Provence Observatory they had detected the first exoplanet, orbiting 51 Pegasi, a solar-type star. It should be noted that three years earlier, researchers Alex Wolszczan and Dale Frail had discovered a triple planetary system around the pulsar PS 1257+12. Theirs was a different

Velocimetry

The procedure known as velocimetry enables us to indirectly detect the presence of an extrasolar planet orbiting a star by revealing the gravitational perturbations induced by that planet. The presence of a companion body near a star causes the star to exhibit a small movement around the common centre of gravity, with a period equal to the companion's period of revolution. With reference to their common centre of gravity, the orbital radii of the star and its companion are in inverse proportion to their masses. It may be assumed that this movement around the centre of gravity is circular. The movement of the companion cannot be measured, as it is not bright enough to be detected. However, the star's movement can de detected, in projection along the line of sight, by measuring its radial velocity. This is determined by high-resolution spectroscopy in the visible region, whereby the measurement of the displacement of a large number of lines gives the radial velocity of the star to an accuracy better than 10 m/s. From this can be deduced the quantity $M \sin i$ (i being the angle of inclination of the system in relation to the plane of the sky) and the orbital period P of the companion.

Velocimetry has proved to be more accurate than direct astrometry, which involves precise measurements of the position of a star compared with the positions of nearby stars. By this method, the first exoplanet was detected in 1995, orbiting the solar-type star 51 Pegasi, and all discoveries of such planets since then have been made in the same way.

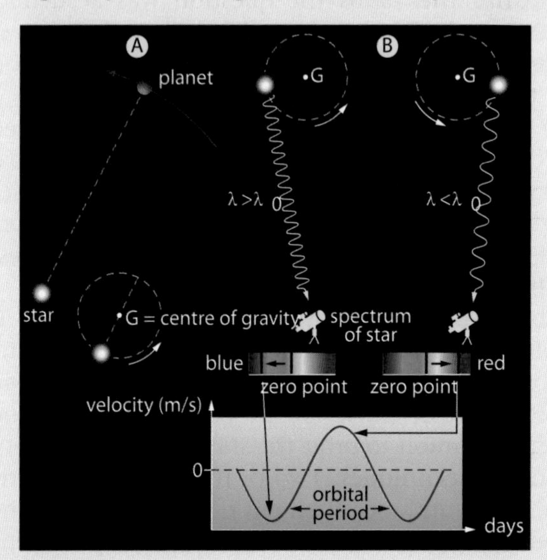

Figure 8.3. The velocimetric method for the detection of exoplanets. (a) The planet orbits its star, but in reality the planet and the star revolve around a common centre of gravity G. (b) Using the Doppler–Fizeau effect an observer can detect the movements of the star, which are normally induced by a sufficiently massive planet or a set of planets. The successive redshifts and blueshifts – characteristic of stars influenced by the presence of an orbiting companion – appear on a graph as a sinusoidal curve.

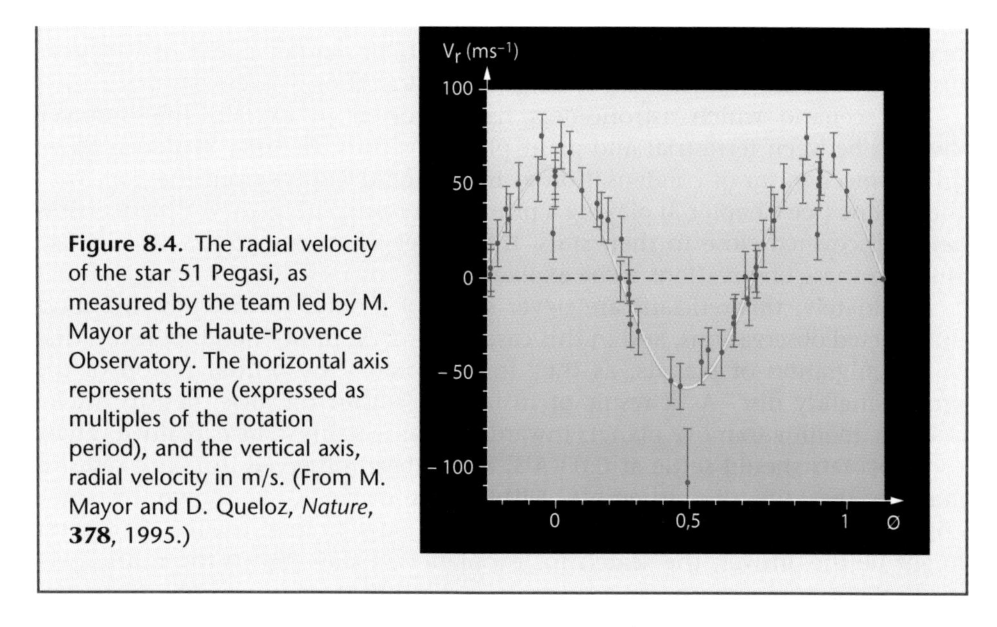

Figure 8.4. The radial velocity of the star 51 Pegasi, as measured by the team led by M. Mayor at the Haute-Provence Observatory. The horizontal axis represents time (expressed as multiples of the rotation period), and the vertical axis, radial velocity in m/s. (From M. Mayor and D. Queloz, *Nature*, **378**, 1995.)

technique, based on the analysis of the very faint shifts in periodicity, observed in the pulsar's periodic radio signal. However, this discovery had not excited quite as much interest within the scientific community. Pulsars – remnants of supernova explosions – are extremely dense stars in ultra-rapid rotation, completely unlike anything in the solar system.

The discovery made by the Geneva observers spurred others to action, and an American team led by Geoffrey Marcy, having confirmed Mayor and Queloz's result, set out to find other exoplanets. The two groups of researchers, together with other teams, were soon scouring the sky, and by 2003 more than a hundred exoplanets had been detected around nearby stars, mostly of solar type.

GIANT EXOPLANETS NEAR STARS

What are the principal characteristics of the exoplanets so far detected? They are nearly all massive, equalling or surpassing Jupiter in size. This is no surprise, as the technique used is all the more successful if the companion body is of high mass, since it must be massive enough to perturb the motion of its parent star. To again cite Jupiter as an example, its presence causes a motion of the Sun of 13 m/s in relation to its environment – a value which has to be within the capability of instruments observing nearby stars. Velocimetric techniques had attained the 10-m/s target by the end of the 1990s. So, 'Jupiters' must be the object of the search, in what is known as an 'observational bias'. A startling fact then began to emerge. Astronomers had found a large number of 'exo-Jupiters' – even more than they had initially expected – but most of these giant exoplanets turned out to be in orbits quite close to their central stars. As was the case with 51 Pegasi, many of them orbited at around 0.05 AU from their star (an eighth of Mercury's distance from the Sun), corresponding to a period of only a few days!

The scenario which astronomers have evolved to explain the observed division between terrestrial and giant planets within the solar system is based on the mechanism of condensation at increasing distances from the Sun, with the ice line (see Chapter 3) playing a primordial role. With giant exoplanets now being discovered close to their stars, the model seems completely discredited. How are these observations to be explained?

Fortunately, theoreticians are never short of models to interpret the most unexpected observations, and in this case the process most often invoked is that of the migration of planets, as they form, towards the central region of the protoplanetary disc. As a result of instability within the disc, density waves develop, tending to move planets inwards. What remains to be explained is why those planets should settle at 0.03 AU, as has been observed in many cases. It may be that the disc is empty within this distance, as shown by some observations of protoplanetary discs around stars in our Galaxy. Whatever might be the answer, the search for exoplanets, today and in the future, will continue to inspire many research projects, both theoretical and observational.

ARE THERE OTHER EARTH-LIKE PLANETS?

It is difficult to imagine that giant exoplanets, like those already detected, could be favourable sites for the emergence of life. These planets are most probably gas giants resembling Jupiter or the other large planets of the outer solar system. No exoplanets have yet been found with temperatures and pressures comparable to those of Earth, but the techniques needed to find them are currently being put in place.

As for the maximum size of an exoEarth: we could estimate it by looking again at the criterion involving the difference in mass between the terrestrial and the giant planets. The critical value marking the separation between the two groups is 10–15 Earth masses. Beyond this value, the gravitational field of the planetary nucleus brings about the inward collapse of the surrounding gas, creating a gas giant. The maximum radius for an exoEarth must therefore be of the order of 2.5–3.5 Earth radii, according to whether the nucleus is rocky (with a density of about 5 g/cm^3) or icy (2 g/cm^3).

How can we detect exoEarths? Using current velocimetric techniques, it will undoubtedly be difficult to measure shifts in velocity of less than a few m/s. The task is to overcome the effect of possible fluctuations inherent in the radiation of the star. There is another, more promising, method which has already borne fruit in the detection of giant exoplanets: the observation of the transits of exoplanets in front of nearby stars. The technique is simple. If a nearby star possesses a planet, and the observer is located in its orbital plane, then the moving planet will appear to cross the face of the star. As this happens, the observer will detect a slight diminution in the brightness of the star as the planet occults a small part of

Figure 8.5. The French COROT space mission, to be launched in 2006 by CNES. (Courtesy Centre National d'Etudes Spatiales.)

the star's disc. The phenomenon is analogous to an eclipse of the Sun, with the Moon playing the part of the planet (see fact box, p. 174), or to the less frequent transits of Mercury and Venus across the Sun. The last transit of Venus took place on 8 June 2004.

In the case of Jupiter, with a diameter ten times less than that of the Sun, the diminution in brightness would be 1%, and in the case of the Earth, 0.01%. It has been possible to observe a transit involving a giant planet from Earth, but to achieve the necessary stability for observing transits by Earth-sized exoplanets, a space platform is needed. This is the objective of the French COROT space mission (scheduled for launch in October 2006), carrying a battery of cameras to be aimed at various star-fields during a period of several months (Figure 8.5). Its continuous observation of thousands of stars may lead to the first discovery of Earth-like exoplanets.

POSSIBILITIES OF LIFE ON EARTH-LIKE PLANETS

If such planets are found (we know that they exist, but not how many there are), the first question to arise will be that of the possibility of life existing there. With no other example upon which to base our thinking, we will have to refer to known conditions on Earth, and seek out liquid water, which has played a pivotal and even indispensable role in the emergence of life on our planet. As has already been stated, without it there is no possibility of the development of amino acids – the basis of that life.

The planetary transit method

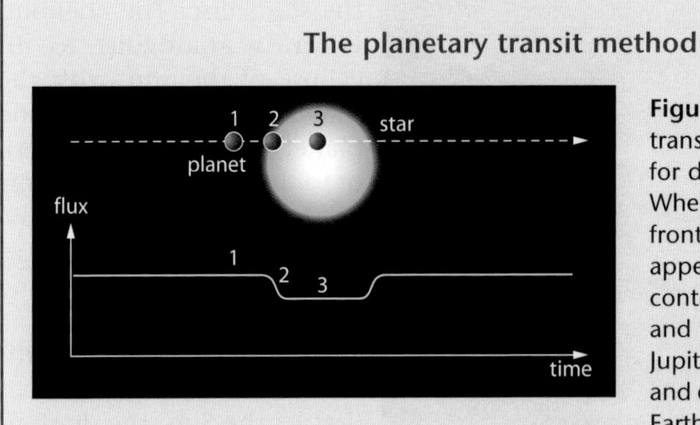

Figure 8.6. The planetary transit or occultation method for detecting exoplanets. When the planet passes in front of the star, the star appears to dim slightly. The contrast between fluxes 1, 2 and 3 is, in the case of a Jupiter-sized exoplanet, 10^{-2}, and of the order of 10^{-4} for an Earth-sized planet orbiting a solar-type star.

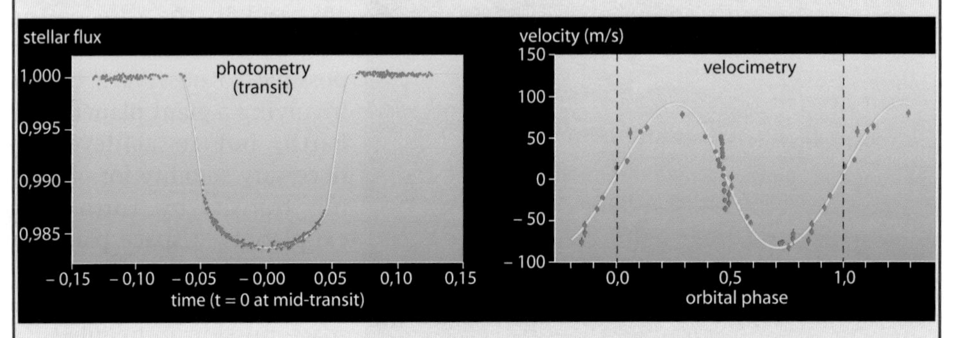

Figure 8.7. The detection of exoplanet HD209458b by occultation and velocimetric methods.

It may happen that a planet orbiting its star will pass between that star and the Earth. As the planet crosses the face of the star there will be a diminution in the apparent brightness of the star. If the radii of the star and the exoplanet are respectively R and r, the relative drop in brightness will be equal to $(r/R)^2$. Taking the example of the solar system as viewed from outside it, Jupiter's passage across the Sun would cause an apparent decrease in the brightness of the Sun by 1%. In the case of the Earth, with a radius only 0.01 that of the Sun, the diminution would be only 0.01%.

It should be noted that the transit is observable only if the geometrical configuration is favourable; that is, if the angle between the plane of the orbit and the line of sight is less than R/a, where a is the distance of the exoplanet from the star. In the case of a planet like Earth, the probability is 0.5%. To achieve success, this method involves the observation of entire star-fields for weeks or even months.

Repeated observations from the ground can detect decreases of the order of 1% in the brightness of a star, thus revealing the presence of giant exoplanets. Such a body was detected orbiting the star HD209458 – an observation confirmed by the Hubble Space Telescope. From Earth, however, photometric precision cannot surpass 0.01%, and space-based detection of exoplanets is therefore necessary. This is the mission objective of COROT, now being developed at the Centre National d'Etudes Spatiales (CNES), and of later missions such as Kepler (NASA).

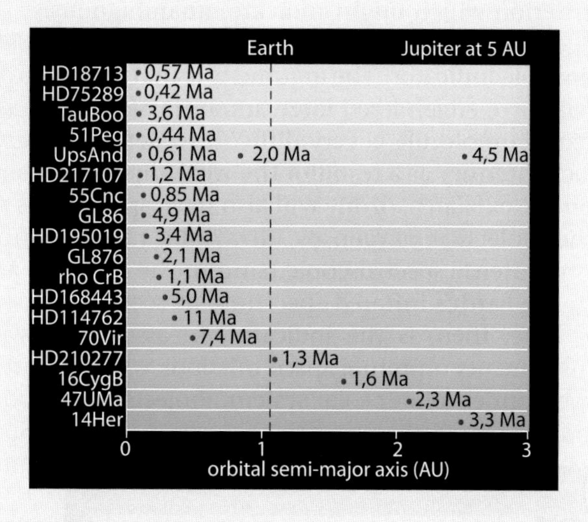

Figure 8.8. A list of the first exoplanets to be discovered, in order of their semi-major axes (distances from their star). More than 130 exoplanets are now known, and the list continues to grow. The pattern shown in the figure is confirmed. A large number of giant exoplanets are quite near their stars, implying a formation scenario different from that which operated in the solar system.

We must therefore evaluate, for any given star, the distance from it at which liquid water may be found – the *habitability zone*. The two factors involved are the mass of the star and its spectral type (directly related to its surface temperature), both of which determine the amount of energy radiated. The habitability zone (located around a distance of 1 AU in the case of the Sun) is further from the star, the greater its mass and temperature. The distance of this zone could be from 0.03 AU outwards to more than 10 AU, depending on the nature of the star. Note that this is not a hard and fast notion. It is based upon the temperature at which one might hope to find liquid water, bearing in mind the amount of energy emitted by the central star. Remember that, in the case of the Earth, a modest greenhouse effect raises its surface temperature by about 30°, and the same effect might come into play on exoplanets. Moreover, within the solar system the Earth may not be unique in harbouring, or having harboured, liquid water. We have already touched upon the notion of ancient seas on Mars; and then there is that ocean which might exist below the surface of Europa.

HOW DO WE FIND EXTRATERRESTRIAL LIFE?

Let us imagine that we have at last discovered one or more exoplanets – perhaps with COROT or with other future space missions dedicated to the search. Further imagine that one or more of them is situated within the habitability zone defined above. How would we know if life were present there? The spectral signature of water vapour will not suffice to determine the additional presence of liquid water, its signature masked by those of the gas. Available spectra of exoplanets and brown dwarfs (similar objects, but very far from their stars, or even isolated) show the presence of water vapour. Can we define a spectroscopic criterion which might indicate, unambiguously, that life has appeared?

After much deliberation, astronomers have settled upon ozone (O_3) as the possible indicator. The argument is that if life has developed upon an exoEarth, it will have engendered large amounts of oxygen, as has happened on Earth. The oxygen molecule is very difficult to detect, because its spectroscopic transitions lack intensity as a result of the weak dipole moment. However, if there is oxygen on the exoEarth, ozone will also have appeared as a result of photodissociation of the molecular oxygen by ultraviolet radiation from the star. Ozone exhibits a very intense spectroscopic band in the infrared, at a wavelength around 10 μm. In the Earth's infrared spectrum, as seen by spacecraft, ozone rules (Figure 8.9)!

Here, then, is the indicator, in association with H_2O, CO_2 or CH_4, which instruments on future space missions will seek out in their search for life beyond the frontiers of our solar system. Projects on the books of NASA and the European Space Agency are particularly ambitious. The Darwin flotilla (Figure 8.10) of six telescopes, of diameters greater than 1 metre, will be launched into an orbit more than 5 AU from the Sun. At this distance they will be unaffected by the 'zodiacal cloud' of interplanetary dust strewn along the plane of the ecliptic. The light of a star falling simultaneously upon the six telescopes will be translated, by interferometry, into an image of that star and its companion. A spectrometer will then register the infrared spectrum of the exoplanet between wavelengths of 7 and 16 μm – a domain featuring not only the signature of ozone but also the signatures of of water, methane, carbon dioxide and ammonia. This project is, to be sure, a long-term one involving enormous technological challenges, and is unlikely to be realised within the next ten years.

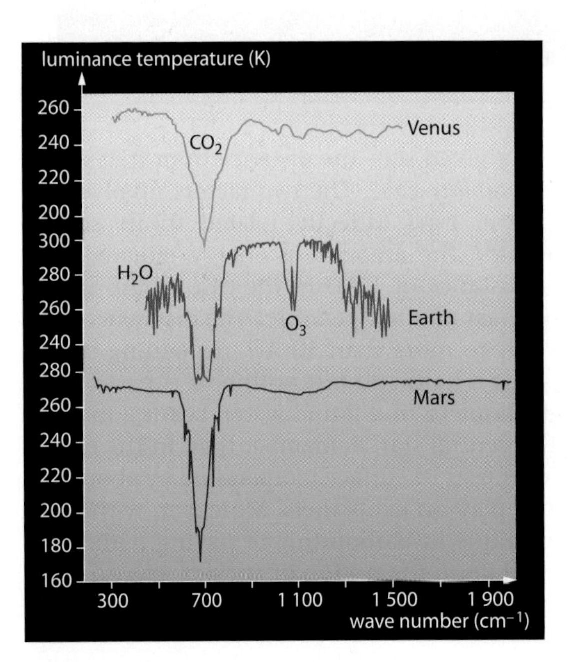

Figure 8.9. These infrared planetary spectra (between 6 and 17 μm) show the signatures of various gases present in their atmospheres: H_2O and CO_2 for terrestrial planets, and O_3 for Earth. The (vertical) luminance scale refers to the flux emitted by the planet. (From R. Hanel *et al.*, 1992.)

Figure 8.10. An artist's impression of the Darwin mission, consisting of a flotilla of six telescopes – each with an aperture of more than 1 metre – in an orbit beyond 5 AU from the Sun. Working as an interferometer, they will be able to image exoplanets in the vicinity of nearby stars.

THE SEARCH FOR EXTRATERRESTRIAL CIVILISATIONS

Let us continue to dream. Suppose that in a decade or two we detect an Earth-like exoplanet displaying the signature of ozone. What then? We will not know at what stage any life on this planet might be, if indeed it exists there at all. And

Figure 8.11. The spherical array of the great Nançay (Sologne) radio telescope, which is used for both transmission and reception in the search for extraterrestrial civilisations. (Courtesy J. Bérezné, CNRS.)

how could we communicate with possible extraterrestrial civilisations? Radio astronomers have been addressing this question for many years, and have had the idea of broadcasting a continuous radio signal at a frequency of 1,420 MHz (corresponding to a wavelength of 21 cm). This is a frequency characteristic of a transition of the hydrogen atoms omnipresent in the Universe (see Chapter 2). It seems likely that other advanced civilisations deeply interested in astronomy will also use such a signal...

Research goes on within the context of project SETI (Search for Extraterrestrial Intelligence), initially funded by NASA. SETI has yet to bear fruit, and continues with private backing, symbolic of the growing collective interest in a question which involves the very core of our identity.

Glossary

Accretion The process by which matter falls together to form solid bodies such as planetesimals, cometary nuclei, asteroids, planets, and so on.

Albedo The fraction of incident light reflected by a body.

Aphelion In an orbit, the furthest point from the Sun.

Arcminute A unit of angular measurement: $1^\circ = 60$ arcmin (minutes of arc), and 1 arcmin = 60 arcsec (seconds of arc). The angular diameter of the Moon is about 30 arcmin.

Asteroid belt The main belt of asteroids, between the orbits of Mars and Jupiter. Asteroids (or minor planets) are solar system objects orbiting the Sun, of sizes between a fraction of a kilometre and nearly 1,000 km. Most of them are in the asteroid belt.

Astronomical Unit (AU) The mean distance between the Earth and the Sun – about 150 million km. This unit is used to measure distances within the solar system.

Black body A body which absorbs all incident radiation (albedo = 0).

Centaur An asteroid travelling between the paths of Jupiter and Neptune, often with a very eccentric orbit. One of them, Chiron, has shown comet-like activity.

Circumstellar region The immediate environment of a star within which its radiation and stellar wind interact with the rest of its protoplanetary disc and the nearby interstellar medium.

Clathrates Solid matrices of water molecules able to entrap another element, for example a molecule of methane or carbon dioxide, or an atom of argon. Clathrates of methane and carbon dioxide exist on Earth. In the case of methane, a molecule of CH_4 is trapped within a matrix composed of about six water molecules. It is possible that clathrates may have played an important part in the formation of the planets and satellites of the outer solar system, and of comets.

Coma A cometary envelope containing gas, ices and dust ejected from the comet's nucleus as a result of solar radiation.

Comet Comets are objects formed from ice and dust, and are a few kilometres in diameter. When they approach the Sun, outgassing of volatile elements causes a coma and a tail to develop, sometimes offering a spectacular sight in the sky.

The difference between comets and asteroids is not very well defined. Certain asteroids may be comets which have lost their reserves of ice after many returns to the vicinity of the Sun.

Cometary nucleus The solid part of a comet, its size ranging from a few kilometres to tens of kilometres, composed of ices and dust grains.

Cometary tail A collection of large cometary dust particles diffusing along the orbit.

Cosmic rays Very energetic particles emanating from the Sun and other stars in our Galaxy.

Differentiation In the solar system, differentiated bodies are those which have undergone internal transformations since they formed. Within the Earth, for example, heating has caused heavy elements to collect at the centre to form a metallic core, leaving a lighter crust above. Low-mass objects such as comets, consisting of conglomerated dust and ice, are not differentiated. Asteroids may show differentiation due to the energy liberated by radioactive elements within them.

Eccentricity A parameter characterising the shape of an orbit. The eccentricity of a circular orbit is 0; of an ellipse, 0–1; of a parabola, 1; and of an hyperbola, >1.

Ecliptic The plane of the ecliptic is the plane in which the Earth travels around the Sun. The orbits of all the planets apart from Pluto are close to the ecliptic.

Exoplanet (extrasolar planet) A planet orbiting a star other than the Sun, and the target of many astronomers! More than 150 exoplanets are now known.

Giant planets Jupiter, Saturn, Uranus and Neptune. They are very massive and not very dense, and their surfaces are hidden by atmospheres composed of hydrogen, helium and, in lesser quantities, methane and ammonia. They have ring systems and many satellites.

Heliocentric distance Distance from the Sun, usually expressed in Astronomical Units (AU).

Ice line Defined by the condensation temperature of water and therefore as a certain distance from the Sun, marking the boundary between the gaseous component of protosolar molecules associated with C, N and O (to the Sunward side of the boundary) and their solid component (beyond the boundary). It has played a major part, among other factors, in the separation of the planets into two distinct classes: terrestrials and gas giants.

Infrared (IR) Spectral domain of wavelengths from 100 μm to 800 nm. With the exception of a few windows accessible from high-altitude observatories, infrared radiation is studied using aircraft, balloons and satellites.

Interstellar grains Small solid particles present in clouds in interstellar space, sometimes incorporated intact into comets.

Isotope An atom of a given element having the normal number of protons and electrons (and therefore the same chemical properties), but a different number of neutrons. The isotopic ratios of an element provide an indication of the history of its formation in the Universe. Some isotopes are unstable and disintegrate over a period which may be very long (sometimes billions of years) or very short, and they can therefore be used as ‘clocks’. The isotopic composition of meteorites has provided information on the age of the solar system, formed 4.55 billion years ago (give or take 100 million years).

Jet The inner coma of a comet is rarely symmetrical. The gas sublimated from the icy nucleus, and the dust accompanying it, emerge from active regions warmed by the Sun, forming jets.

Kelvin A measurement of temperature. Absolute zero is 0 K (–273°.15 C).

Kuiper Belt A collection of planetesimals (asteroids and cometary nuclei) orbiting beyond Neptune, at distances between 30 and 100 AU. The Kuiper Belt is thought to be the source of short-period comets of the Jupiter family.

Light-year The distance light travels in a vacuum in one year (63,000 AU).

Magnetosphere An ionised envelope around the Earth – the result of the interaction of its magnetic field with the solar wind.

Meteor A luminous phenomenon caused by the passage of a meteoroid through the atmosphere; commonly referred to as a 'shooting star'. A group of meteors is known as a shower.

Meteorite An extraterrestrial object which has reached the Earth's surface, having survived its passage through the atmosphere.

Minute of arc (See *arcminute.*)

Nucleosynthesis The train of nuclear reactions leading to the production of the different chemical elements. At the time of the primordial nucleosynthesis, completed a few minutes after the Big Bang, the lightest elements – hydrogen, deuterium, helium and lithium – were created. The heavier elements are synthesised by stars. In stellar nucleosynthesis, hydrogen is transmuted into helium, and so on to form carbon, nitrogen, oxygen, and other heavier elements.

Occultation A phenomenon involving a celestial object (a planet or satellite) passing directly in front of another. When a planet occults a star, the way in which the star's light is absorbed by the planet's atmosphere (if present) tells us about the composition of the atmosphere and the temperature profile according to altitude. If the planet has a ring system that passes in front of the star, the light of the star will appear to dim intermittently. The rings of Uranus and Neptune were discovered in this way.

Oort Cloud A hypothetical cloud of comets uniformly spread at a distance of 20,000 to 100,000 AU from the Sun.

Organic compound Molecules composed of carbon, hydrogen and (possibly) oxygen and nitrogen.

Outgassing The ejection of gas from a celestial body. In the case of the giant planets, outgassing is associated with sublimation of ices from the core due to heating.

Parsec (pc) The distance from the Earth to a star at which the semi-major axis of Earth's orbit subtends an angle of 1 arcsecond; 1 pc = 3.261633 ly.

Perihelion In an orbit, the nearest point to the Sun.

Period The time taken for a body to describe one complete orbit.

Photochemistry The range of chemical reactions caused by solar radiation in an atmosphere.

Photodissociation The dissociation of a molecule due to radiation. In the case of a planetary atmosphere, photodissociation is caused by solar radiation.

Photolysis Chemical decomposition due to light.

Photometry A technique for measuring the intensity of radiation.

Planetesimal When the primordial nebula collapsed into a disc, collisions between dust grains within the interstellar cloud created tiny aggregates about 1 μm across: planetesimals. The next stage saw planetesimals growing as a result of mutual collisions, and then drawing in material from around them (accretion) to become protoplanets.

Prebiotic Prebiotic molecules are organic molecules essential to the chains of reactions leading to the formation of amino acids, constituents of proteins. Some of these molecules are present in the atmosphere of Titan, causing some astronomers to see it as a laboratory of prebiotic chemistry.

Primordial solar nebula (protoplanetary cloud) A cloud of gas and dust which, having collapsed into a disc, gives birth to the Sun and its planets.

Radio The spectral domain of wavelengths greater than 100 μm. It embraces, in particular, the millimetre (1–10 mm) and submillimetre (0.1–1 mm) ranges, parts of which are inaccessible to Earth-based radio telescopes.

Refractory (molecules) Molecules able to remain solid at quite high temperatures; for example, silicates and metals.

Roche limit The distance within which a satellite is destroyed by differential tidal effects on the sides turned towards and away from the planet, overcoming the satellite's cohesion. For a satellite of approximately the same density as its primary, this distance is about 2.5 times the planetary radius. Within this limit we do not expect to find any stable satellites. This is the domain of the rings.

Satellite A celestial body (for example, the Moon) orbiting a planet. Giant planets possess many satellites.

Solar wind; stellar wind A continuous stream of energetic particles flowing from the Sun and other stars. The solar wind is essentially composed of electrons and protons reaching speeds of about 400 km/s by the time they reach the Earth.

Spectroscopy The analysis of the electromagnetic radiation of a star as a function of wavelength. This astrophysical method can be used to identify the elements present in a star, by detection of emissions at certain wavelengths characteristic of these elements.

Sublimation A change of state from the solid to the gaseous.

Terrestrial planet Mercury, Venus, Earth and Mars. They are dense, relatively small, and have atmospheres (except Mercury) and surfaces accessible to observation. They have few or no satellites.

Transit (planetary) Used to detect an exoplanet. In its movement around a star, a planet may pass between the star and the Earth. The passage of the exoplanet in front of the star causes a diminution of the star's brightness. If the star and the planet have radii respectively R and r, the relative diminution of the flux will equal $(r/R)^2$.

Trojans A family of asteroids in the same orbit as Jupiter, at approximately 60° preceding and following the planet.

Ultraviolet (UV) Spectral domain of wavelengths below 350 nm. UV radiation shorter than 300 nm may be observed from the ground, although it is attenuated by the atmosphere. Beyond this wavelength, observation by rockets and satellites is necessary.

Velocimetry A method of indirectly detecting the presence of an extrasolar planet orbiting a star by measuring the effect of gravitational perturbations affecting the radial velocity of the star.

Visible Spectral domain of wavelengths between 350 and 800 nm, corresponding to what is seen by the human eye.

Volatile (molecule) A molecule which sublimates or condenses at a relatively low temperature; for example, molecules which constitute cometary ices.

Bibliography

C. Allègre, *De la Pierre à l'Étoile*, Fayard, 1985.

S.K. Atreya, *Atmospheres and Ionospheres of the Outer Planets and their Satellites*, Springer-Verlag, 1986.

E.M. Antoniadi, *La Planète Mars*, Burillier, 1930.

J. Audouze and G. Israël, *Grand Atlas de l'Astronomie*, Encyclopedia Universalis, 1993.

W.I. Ausich and N.G. Lane, *Life of the Past*, Prentice Hall (4th edition), 1999.

B. Bézard *et al.*, *Nature*, **345**, 508, 1990.

Ph. Bendjoya, *Collisions dans le Système Solaire*, Belin, 1998.

A.Brahic, *Les Comètes*, Collection 'Que sais-je?', Presses Universitaires de France, 1993.

D. Bockelée-Morvan *et al.*, *Icarus*, **133**, 147, 1998.

A. Boss, *Looking for Earths*, Wiley, 1998.

L. Botinellie *et al.*, *La Terre et l'Univers*, Hachette Éducation, 1993.

A. Carion, *Les Météorites et leurs Impacts*, Éditions Armand Colin, 1993.

F. Casoli *et al.*, *Astron. Astrophys.*, **287**, 716, 1994.

R. Chaboud, *Pleuvra, Pleuvra Pas*, Gallimard Jeunesse, 1994.

M.T. Chaine, M.F. A'Hearn and J. Rahe, *Comparative Planetology with an Earth's Perspective*, Kluwer, 1995.

S. Clark, *Extrasolar Planets*, Praxis–Wiley, 1998.

M. Combes *et al.*, *Icarus*, **76**, 404, 1988.

F. Costard, *La Planète Mars*, Collection 'Que sais-je?', Presses Universitaires de France, 2000.

Ph. de la Cotardière (ed.), *Le Grand Livre du Ciel*, Bordas, 1999.

A. Coustenis and F.W. Taylor, *Titan, the Earth-like Moon*, World Scientific, 1999.

A. Coustenis *et al.*, ESA SP-419, 255, 1997.

P. Cox *et al.*, ESA SP-427, 631, 1999.

J. Crovisier *et al.*, *Science*, **275**, 1904, 1997.

J. Crovisier and Th. Encrenaz, *Les Comètes*, Belin/CNRS-Éditions, 1995.

K. Croswell, *Planet Quest*, 1997.

D.P. Cruikshank *et al.*, *Solar System Ices*, B. Schmitt et al. (éd.), Kluwer, 1998.

D.P. Cruikshank *et al.*, *Ann. Rev. Earth Planet Sci.*, **25**, 243, 1997.

Dictionnaire de l'Astronomie, Encyclopedia Universalis, Albin Michel, 1999.
I. de Pater and J. Lissauer, *Planetary Sciences*, Cambridge University Press, 2001.
J. Davies, *Beyond Pluto*, Cambridge University Press, 2001.
P. Drossart *et al.*, *Icarus*, **49**, 20, 1982.
J. Elliot, *Ann. Rev. Astron. Astropys.*, **17**, 445, 1979.
Th. Encrenaz, *La Planète Géantes*, Belin, 1996.
Th. Encrenaz, *Atmosphères Planétaires. Origine et Évolution*, Belin/CNRS-Éditions, 2000.
Th. Encrenaz, *Le Système Solaire*, Flammarion, 2004.
Th. Encrenaz and P. Cox, *L'Eau dans l'Univers*, C.R. Acad. Sci. Paris, **1**, Series IV, 981, 2000.
Encyclopedia of Astronomy and Astrophysics, Institute of Physics Publishing and Nature Publishing, 2001.
Encyclopédie Scientifique de l'Univers: les Étoiles, le Système Solaire, Bureau des Longitudes, Gauthier-Villard, 1986.
P. Feldman, *Comets*, L. Wilkening (ed.), University of Arizona Press, 1982.
H. Feuchgruber *et al.*, *Nature*, **389**, 159, 2003.
F. Forget, F. Costard and Ph. Lognonnè, *La Planète Mars. Histoire d'un autre Monde*, Belin, 2003.
Ch. Frankel, *La Mort des Dinosaures: l'Hypothèse Cosmique*, Masson, 1996.
Ch. Frankel, *La Vie sur Mars*, Seuil, 1999.
T. Gehrels (ed.), *Hazards due to Comets and Asteroids*, University of Arizona Press, 1994.
D. Goldsmith, *The Hunt for Life on Mars*, Dutton, 1997.
R. Greenberg and A. Brahe (eds.), *Planetary Rings*, University of Arizona Press, 1984.
R. Hanel *et al.*, *Exploration of the Solar System by Infrared Remote Sensing*, Cambridge University Press, 1992.
G. Hertzberg, *Infrared and Raman Spectra*, Van Nostrand, 1945.
G. Horneck and C. Baumstark-Khan (eds.), *Astrobiology: The Quest for the Conditions of Life*, Springer-Verlag, 2002.
J.T. Houghton, *The Physics of Atmospheres*, Cambridge University Press, 1986.
B.M. Jakosky, *The Search for Life on Other Planets*, Cambridge University Press, 1998.
B.M. Jakosky and C.B. Farmer, *Mars*, University of Arizona Press, 1992.
S. Joussame, *Climats, d'hier à Demain*, CEA/CNRS-Éditions, 1993.
J.F. Kerridge and M.S. Matthews (eds.), *Meteorites and the Early Solar System*, University of Arizona Press, 1992.
H.H. Kieffer, B.M. Jakosky and C.W. Snyder, *Mars*, University of Arizona Press, 1992.
V.A. Krasnopolsky, *Photochemistry of the Atmospheres of Mars and Venus*, Springer-Verlag, 1986.
K.B. Krauskopf and A. Beiser, *The Physical Universe*, McGraw-Hill, 1979.
K.R. Lang and Ch. A. Whitney, *Vagabonds de l'Espace*, Springer-Verlag, 1993.
H. Larson, *Ann. Rev. Astron. Astrophys.*, **18**, 43, 1980.

H. Larson *et al.*, *Astrophys. J.*, **338**, 1106, 1989.
J.S. Lewis, *Physics and Chemistry of the Solar System*, Academic Press, 1995.
P. Léna (ed.), *Les Sciences du Ciel*, Flammarion, 1996.
P. Léna, *Méthodes Physiques de l'Observation*, Inter-Éditions du CNRS, 1986.
A.-C. Levasseur-Regoud, *Les Comètes et les Astéroides*, Éditions du Seuil, 1997.
J. Lunine, *Earth: Evolution of a Habitable World*, Cambridge University Press, 1999.
M. Mayor and P.-Y. Frei, *Les Nouveaux Mondes du Cosmos*, Éditions du Seuil, 2001.
M. Mayor and D. Queloz, *Nature*, **378**, 355, 1995.
Ph. Miné, *Bizarre Big Bang. L'Épopée de la Physique*, Belin-Pour la Science, 2001.
D. Morrisson (ed.), *Satellites of Jupiter*, University of Arizona Press, 1981.
D. Morrisson and T. Owen, *The Planetary System*, Addison-Wesley, 1996.
E. Nesme-Ribes and G. Thuillier, *Histoire Solaire et Climatique*, Belin, 2000.
K.S. Noll, H.A. Weaver and P.D. Feldman, *The Collision of Comet Shoemaker-Levy 9 with Jupiter*, Cambridge University Press, 1996.
B. O'Leary, J.K. Beatty and A. Chaikin (eds.), *The New Solar System*, Cambridge University Press and Sky Publishing Corporation, 1999.
T. Oka, *Science*, **277**, 328, 1997.
T. Owen *et al.*, *Science*, **240**, 1767, 1988.
K.O. Pope *et al.*, *Earth, Moon and Planets*, **63**, 93, 1993.
R. Prinn and T. Owen, *Jupiter*, T. Gehrel (ed.), University of Arizona Press, 1976.
S.I. Rasool, *Système Terre*, Collection Dominos, Flammarion, 2001.
J.H. Rogers, *The Giant Planet Jupiter*, Cambridge University Press, 1995.
M. Seguin and B. Villeneuve, *Astronomie et Astrophysique*, Éditions du Renouveau Pédagogique, 1995.
J.H. Shirley and R.W. Fairbridge, *Encyclopedia of Planetary Sciences*, Chapman and Hall, 1997.
T. Sill and R.N. Clark, *Satellites of Jupiter*, D. Morrison (ed.), University of Arizona Press, 1982.
J.R. Spencer and J. Mitton, *The Great Comet Crash*, Cambridge University Press, 1995.
S.R. Taylor, *Solar System Evolution. A New Perspective*, Cambridge University Press, 1992.
T. Tsuji *et al.*, ESA-SP427, 229, 1999.
R. Trompette, La Terre. *Une Planète Singulière*, Belin-Pour la Science, 2003.
X. Van Dishoek *et al.*, ESA-SP427, 579, 1999.
P.R. Weissman, L.-A. McFadden and T.V. Johnson (eds.), *Encyclopedia of the Solar System*, Academic Press, 1999.
B. Zanda, M. Rotaru and Ph. de la Cotardière (eds.), *Les Météorites*, Bordas, 1996.

Index

Printing: Mercedes-Druck, Berlin
Binding: Stein+Lehmann, Berlin